AF575657

Copyright © 2024 by David Doyle

Library of Congress Control Number: 2023941003

All rights reserved. No part of this work may be reproduced or used in any form or by any means—graphic, electronic, or mechanical, including photocopying or information storage and retrieval systems—without written permission from the publisher.

The scanning, uploading, and distribution of this book or any part thereof via the Internet or any other means without the permission of the publisher is illegal and punishable by law. Please purchase only authorized editions and do not participate in or encourage the electronic piracy of copyrighted materials.

"Schiffer Military" and the arrow logo are trademarks of Schiffer Publishing, Ltd.

Designed by Christopher Bower
Cover design by Justin Watkinson
Type set in Impact/Minion Pro/Univers LT Std

ISBN: 978-0-7643-6781-6
Printed in India

Published by Schiffer Publishing, Ltd.
4880 Lower Valley Road
Atglen, PA 19310
Phone: (610) 593-1777; Fax: (610) 593-2002
Email: Info@schifferbooks.com
Web: www.schifferbooks.com

For our complete selection of fine books on this and related subjects, please visit our website at www.schifferbooks.com. You may also write for a free catalog.

Schiffer Publishing's titles are available at special discounts for bulk purchases for sales promotions or premiums. Special editions, including personalized covers, corporate imprints, and excerpts, can be created in large quantities for special needs. For more information, contact the publisher.

We are always looking for people to write books on new and related subjects. If you have an idea for a book, please contact us at proposals@schifferbooks.com.

Acknowledgments

This book would not have been possible without a great deal of help from a number of friends and associates, to whom I am deeply indebted. Among these are Tom Kailbourn, Scott Taylor, Chris "Toadman" Hughes, John Charvat, the late Don Moriarty, Dana Bell, Steve Zaloga, Sean Hert, Jim Gilmore, and the staffs of the US National Archives and Records Administration, the TACOM history office, the Rock Island Arsenal Museum, and the former Patton Museum. I owe a special thanks to my wife, Denise, who, in addition to helping locate and scan materials for this and countless other books, remains an ongoing source of encouragement and support.

Contents

Introduction

The M48 Patton, with its distinctive curved cast hull front, was the mainstay of US tank forces through much of the 1950s. Beginning in 1952, nearly 12,000 M48-series tanks were produced, but by the mid-1950s experts began to feel that the M48's 90 mm gun was inadequate, Efforts to produce a heavier-armed tank culminated in the 105 mm armed M60, introduced in 1959, two years after output of the M48 ceased. *National Archives*

The M60 was the principal tank fielded by the US military from 1960 until it was gradually phased out in favor of the Abrams. The M60, while often referred to as a "Patton" tank, was never actually so named, that name instead being bestowed upon its similar-looking predecessor, the M48.

The 90 mm armed M48 itself was relatively new when in 1956 a defector drove a Soviet T-54A onto the grounds of the British embassy in Budapest during the Hungarian Uprising. After examining it, British specialists concluded that their tanks could not defeat the armor of the Soviet vehicle, and they set out to develop a larger tank gun, and this information made its way to US Army planners as well.

The US Army already had a new tank, the T95, on the drawing board, with many advanced features, including a 90 mm cannon. However, the T95 was still well away from being ready for production. As an interim measure, a new tank was engineered, essentially placing a 105 mm gun based on a British design in a modified M48A2 turret, which was in turn placed on a chassis developed from that of the M48A2 as well. Designated XM60, the new tank was standardized as the M60 in March 1959. The hull of an M60 is readily distinguished from that of the M48 series in that the latter has a cast, rounded glacis, while the M60 glacis consists of flat plates that come to a knife edge. This configuration was intended to facilitate production with siliceous-cored armor, although such armor was never used. Standard homogeneous steel armor was used instead.

The M60, through a variety of variants, remained in production into 1987, and though it no longer serves US forces in combat, it remains in use by several nations.

The Soviet T-54 medium tank made its debut in the 1950s. It was simple to operate, made a relatively small target, was produced in large quantities, and had a powerful 100 mm main gun. The threat of the T-54 spurred the United States to develop a tank capable of countering it: the M60. *Steve Zaloga collection*

In the mid-1950s, the Army developed the 90 mm T95 gun tank as the successor to the M48. The first pilot T95, 9B1043, is shown here with turret reversed in February 1958. However, the T95 program was canceled in 1960, by which time development of the M60 was well underway. *TACOM LCMC History Office*

CHAPTER 1

M60

The first production model of the M60 carried that simple designation, without an alphanumeric suffix. It was characterized by a straight bow, unlike the curved bow of the M48 tanks, and an M48-style turret. Here, the M60 with registration number 9B3484 performs a demonstration at the forty-fourth annual meeting of the US Ordnance Association, at Aberdeen Proving Ground, Maryland, on October 4, 1962. An early-type mounting bracket for a searchlight is on the mantlet. *National Archives*

In preparation of placing the new tank in production, in September 1958 Chrysler Corporation was awarded an Advance Production Engineering Contract, which included provisions for the manufacture of four pilot vehicles. On December 11, 1958, Gen. Maxwell Taylor ordered the M60 into production. On March 16, 1959, per Ordnance Committee Minutes (OCM) 37002, the new vehicles were standardized as the "105 mm Gun Full Tracked Combat Tank M60."

Chrysler completed the first of the four pilots in June 1959 and delivered it to Detroit Arsenal on July 3. It was subsequently sent to Aberdeen Proving Ground, arriving on July 24. The second pilot was completed on August 4 and was retained for use in preparing publications. The third pilot, completed on September 2, was also delivered to Detroit Arsenal, which immediately dispatched the vehicle to Fort Knox for user testing. The fourth pilot was delivered to Detroit Arsenal on October 26.

Production of the M60 began at Chrysler's Newark, Delaware, tank plant, which previously had manufactured M48 tanks, the vehicle the plant had been built to produce.

However, with the Korean War brought to an armistice, tank production began to be consolidated, and Chrysler records indicate that after 360 M60s were completed in Newark, M60 operations were moved to the Chrysler-operated Detroit Tank Plant.

The M60 first began to reach troops in December 1960, with units in Germany—the forecast front line against a Soviet threat—receiving the initial shipments.

M60, registration number 9B3090, with a white silhouette of a panther on its 105 mm gun barrel, is symbolic of the first generation of M60 main battle tanks. Over more than three decades, these tanks would see numerous improvements, making it a highly effective fighting machine. *National Archives*

The right side of the M60 pilot, US Army registration number 9B3057, is seen at the Detroit Arsenal on July 1, 1959. The early-style side-loading air cleaner on the fender below the turret bustle had an access door with hinges at the rear and a latch mechanism at the front. *Patton Museum*

The M60 pilot tank is viewed from the left side in a photograph taken on November 19, 1959. "U.S.A. 9B 3057" is painted in small characters on the forward storage box on the fender. The turret was similar to the one used on the M48A2 tank, and the cupola was the M29 type. *National Archives*

The pilot M60 moves toward the firing line during its first public appearance, at the annual meeting of the American Ordnance Association at Aberdeen Proving Ground on October 8, 1959. The vehicle had officially been completed at the Detroit Arsenal on July 3, 1959, and had arrived at Aberdeen Proving Ground shortly thereafter. Unlike the curved front bowline of the M48, the front bowline of the M60 pilot is straight across. *National Archives*

The pilot M60 displays its left side during a demonstration for the American Ordnance Association at Aberdeen Proving Ground, Maryland, on October 8, 1959. In addition to the "0" placard painted on the vehicle, a placard with the number "46" had been affixed to the handrail. *National Archives*

M60 9B3667 is viewed from the right rear. From this angle, the M60 looked very similar to an M48A1 or M48A2 tank, the chief distinguishing feature being the noticeably different styles of the cupolas. The dome below the cupola is the housing for the right rangefinder objective. *National Archives*

M60, registration number 9B3204, was the subject of experiments by the US Army Armor Board at Fort Knox, Kentucky, to equip these tanks with an underwater-fording kit. This kit was designed to allow the tank to ford rivers and bodies of water up to 15 feet in depth. A four-section conning tower attached over the loader's hatch on the turret was a major part of the kit. This M60 is also equipped with a bulldozer blade. *National Archives*

As seen in another photo of the same M60 with underwater-fording kit and bulldozer blade, registration number 9B3204, the headlights and brush guards were extended above the raised dozer blade. For depth reference, feet and inches scales were painted in white on the conning tower. *National Archives*

The M60 with an underwater-fording kit is seen from the right side. Development and testing of this kit occurred in 1962 and 1963. In addition to the conning tower, the kit also included a waterproof mantlet cover, a bilge pump, and other sealant materials. *National Archives*

The conning tower consisted of four sections, stacked one on top of the other and held together with tie-down straps. A removable ladder made up of several sections was placed in the interior of the conning tower for crew entry and egress. Also, steps were available on the exterior. *National Archives*

An M60 on outdoor display exhibits the characteristic straight-across joint between the bottom of the glacis and the top of the lower front plate of the hull. Dustcovers are installed on the cupola and the turret, but the cover over the 105 mm gun shield is frayed at the top, allowing the wire hoop to pop through. On the turret roof to the front of the cupola is the hood for the gunner's periscope, with a cover on the front to protect the periscope when not in use. *Chris Hughes*

The paint on this M60 displayed outdoors is heavily weathered, and rust is evident below the vision blocks on the cupola. The stencil on the side of the air cleaner housing cautions that after a nuclear-biological-chemical (NBC) attack, the cleaner must be serviced by NBC-protection-equipment personnel. The bogie wheels are the forged-aluminum type, with reinforcing gussets around the perimeter. *Chris Hughes*

Details of the cupola and the right side of the turret of a display M60 are presented, including, on the cupola, two of the vision blocks and the commander's periscope housing. Below the cupola is the dome-shaped housing for the right side of the M17C rangefinder. Two spare track shoes are clamped to the handrail on the side of the turret. *Chris Hughes*

Two spare bogie wheels are secured to the baggage rack on the rear of the turret. Multiple 5-gallon liquid containers are stored on holders on the fenders; the armored box to the left of the container on the right side of the vehicle contains the external interphone set. *John Charvat*

To the rear of the air cleaner housing on the right fender is the rear storage box. The method of securing the eyes of the tow cables to the turret is illustrated. Also in view are the rear shock absorber, the right sprocket, and several of the bogie wheels and track-support rollers. *Chris Hughes*

A rear view of an M60 shows the two transmission-access grille-doors. On the right door is a square inset, on which an engine-exhaust trunk for a deepwater-fording kit could be attached. Taillight assemblies are in arch-shaped guards below the access doors. Between the taillights are the tow pintle and two access plates for the transmission. Curved mudguards extend up from the taillight guards to the mudguards on the rears of the fenders. *Chris Hughes*

The engine deck, travel lock for the 105 mm gun barrel, external intercom box, ventilation grille-doors, and the right air cleaner and rear storage box are viewed from above the right side of an M60. On the turret to the far right are two holders for tow-cable eyes. *John Charvat*

On the left-rear corner of the turret roof are the hood and splash guard for the turret ventilator. To the rear of the liquid container on the fender is a triangular support for the fender, with lightening holes of varying sizes. *John Charvat*

An M60, registration number 9B3429, is viewed from the left side. When the tracks were properly tensioned, as seen here, there was little or no sag in their upper runs. *Chris Hughes*

This M60 is painted in three-color NATO camouflage, consisting of green, black, and brown. To the front of the cupola is the deflector for the cupola machine gun. Shaped like an inclined letter "L," this device, which could be folded down, physically prevented the cupola gun from shooting up the vehicle by diverting the barrel of the gun. *Chris Hughes*

An M60 turret is viewed close-up from the left side. The steel pad welded to the side of the cupola below the center vision block was an early-production feature and was for mounting a .50-caliber M2 HB machine gun on the exterior of the cupola. The driver's instrument panel is visible through the open hatch to the left. *Chris Hughes*

The same turret is observed from the left rear, showing a 5-gallon liquid container in a holder, two tow cables, the turret ventilator hood and its splash guard, the loader's open hatch door, and the baggage rack, with folded tarpaulins in it. *Chris Hughes*

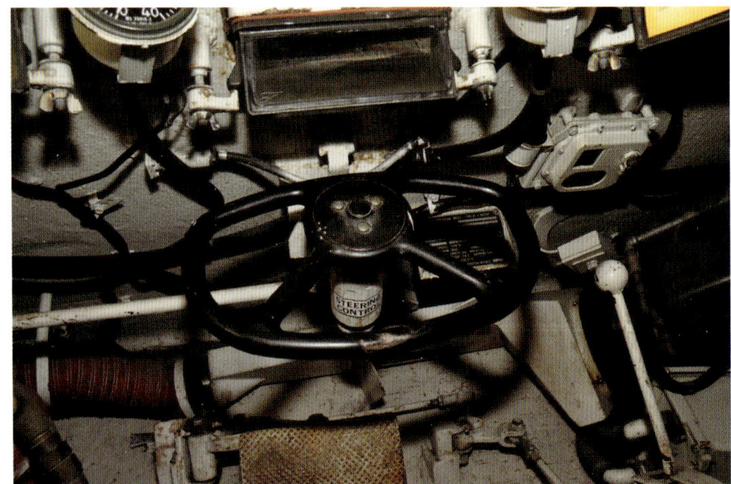

The driver's compartment is viewed facing forward, with the steering wheel at the center, and the driver's center periscope above the wheel. Below the wheel is the brake pedal, and to the right of the wheel is the transmission gearshift lever. The vehicle data plate is to the front of the right side of the steering wheel. *Chris Hughes*

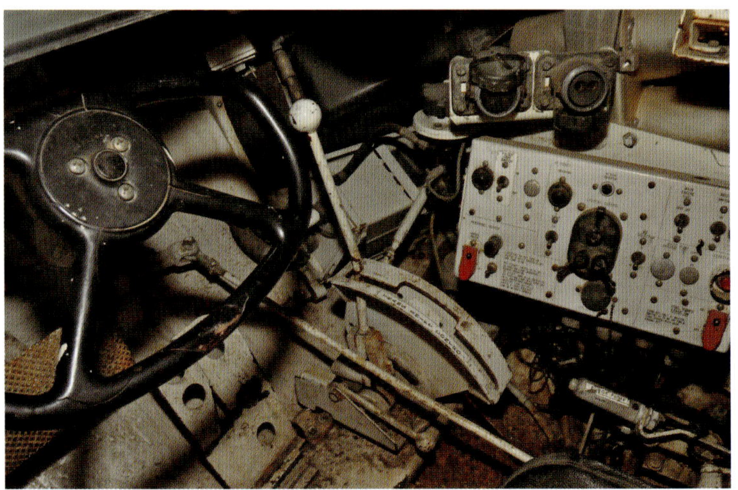

To the right of the steering wheel and the transmission shift is the driver's master control panel. This is the late-type panel, which varied from the early type mainly in that it had a starter switch on the right side. Light-control switches are at the center of the master control panel. *Chris Hughes*

Moving toward the upper right (i.e., rear) of the preceding photo, at the center is the driver's instrument panel, with the master control panel to the lower left and the speedometer/odometer to the far left. *Chris Hughes*

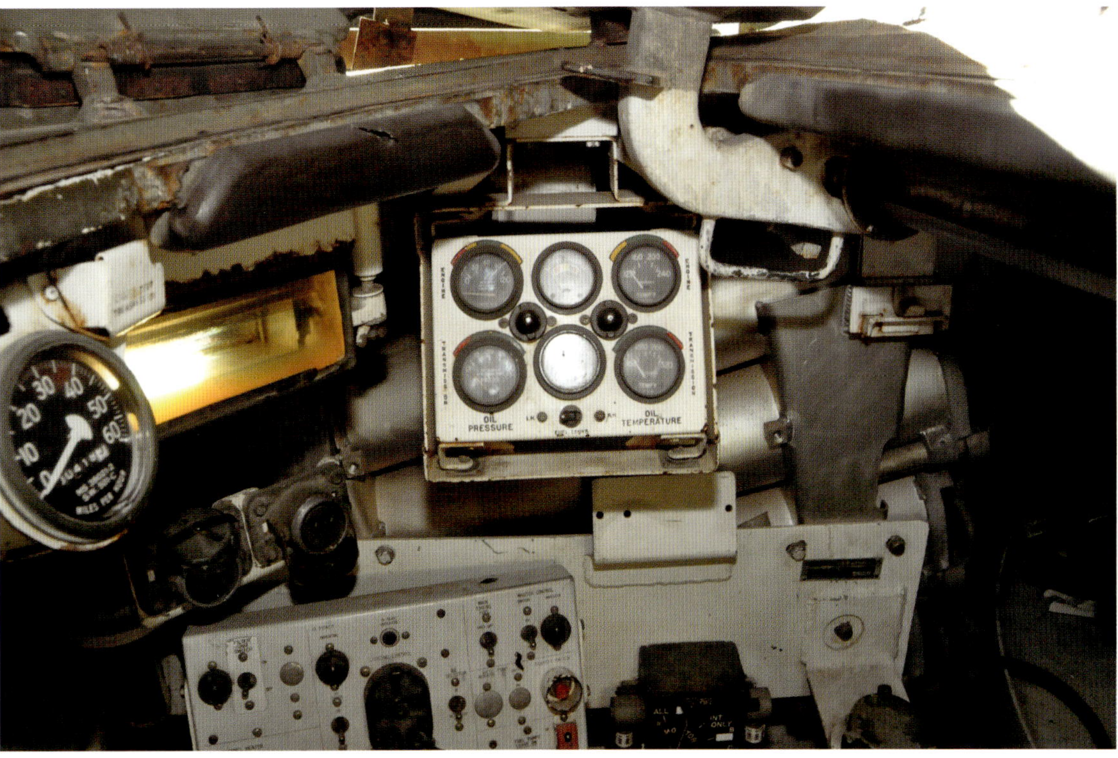

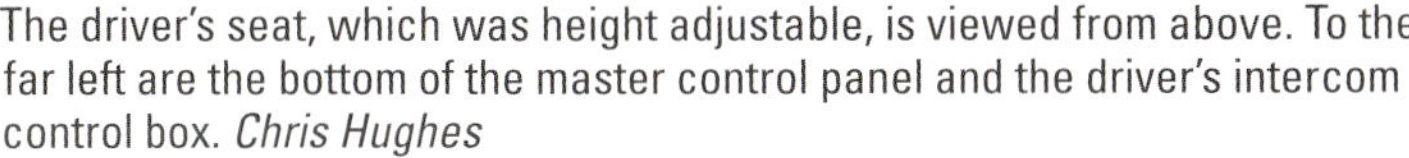

The driver's seat, which was height adjustable, is viewed from above. To the far left are the bottom of the master control panel and the driver's intercom control box. *Chris Hughes*

In the turret of an M60, the breech ring and the breech block of the 105 mm gun are viewed from the rear. Although visible only under high magnification, stamped on the lower part of the breech ring is "GUN, 105 mm, M68 No. 166," along with other markings for Watervliet Arsenal and the Ordnance Corps. *Chris Hughes*

The vehicle commander's station in an M60 is viewed from above the recoil guard of the 105 mm gun. At the top is the cupola. At the center is the commander's control handle, which was linked to the gunner's controls and enabled the commander to take over control of the guns. Running across the upper left of the photo is the M17C rangefinder. On the side wall of the turret, to the rear of the commander's control handle are, *left*, the commander's intercom control box and, *right*, the searchlight control box. *Chris Hughes*

In a view from the gunner's position in the turret, at the top are, *left*, the sighting telescope; *center*, the daylight viewer of the gunner's periscope; and *right*, the nighttime (infrared) viewer of the periscope. Immediately below these viewers is the gunner's selector switch box. At the bottom are the gunner's control handles. Above those controls is the handle for manually traversing the turret. To the lower right is the ballistic computer. *Chris Hughes*

The ballistic computer is shown more fully. To the immediate left of the ballistic computer is the azimuth indicator. *Chris Hughes*

In the ready rack for 105 mm ammunition on the left side of the turret basket, HEP-T high-explosive squash-head shells are stored in the foreground, and armor-piercing discarding-sabot (APDS) rounds, with their thin, pointed projectiles, are in the background. To the lower right are tubes for storing ammunition. *Chris Hughes*

DEVELOPMENT

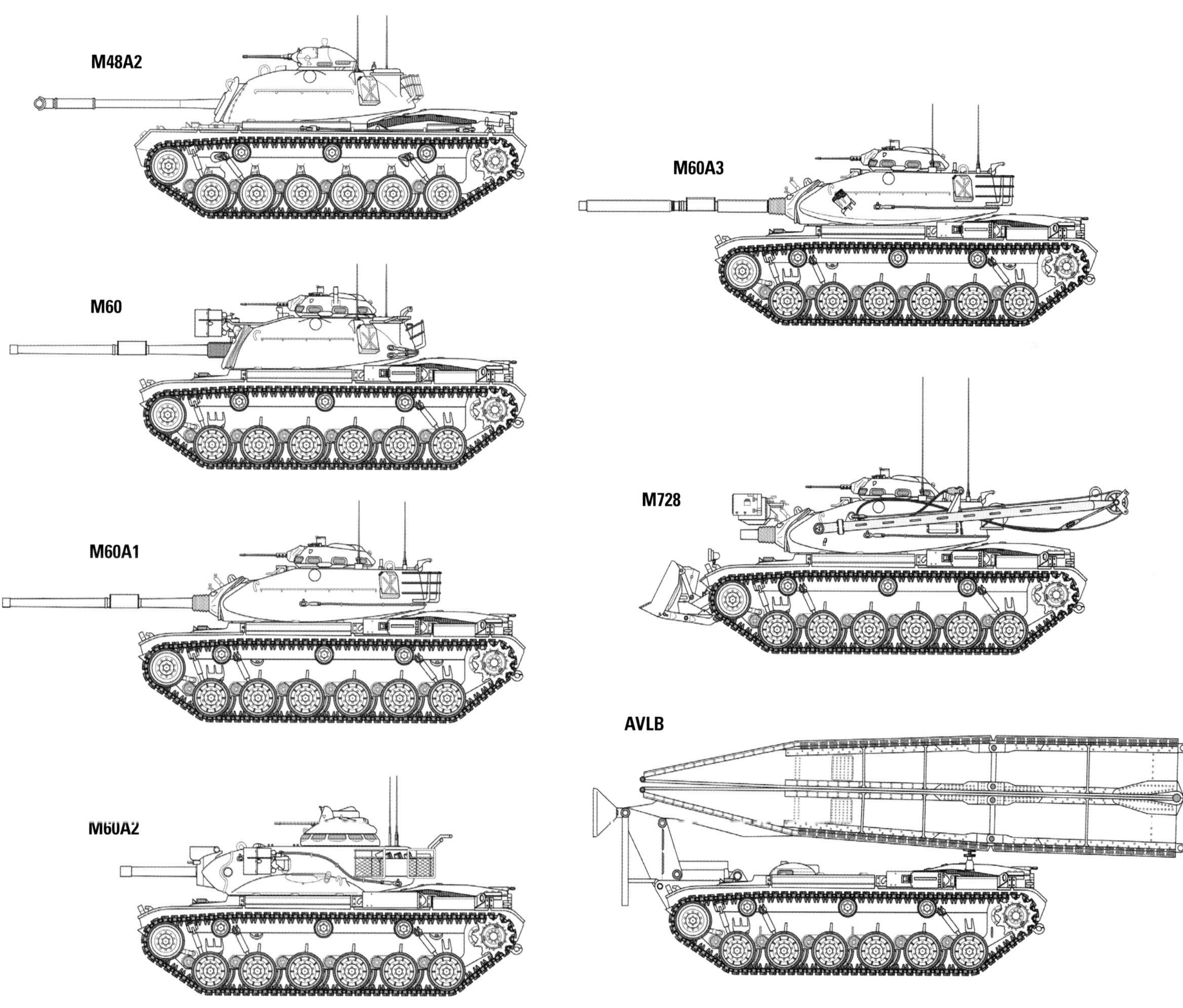

CHAPTER 1
M60A1

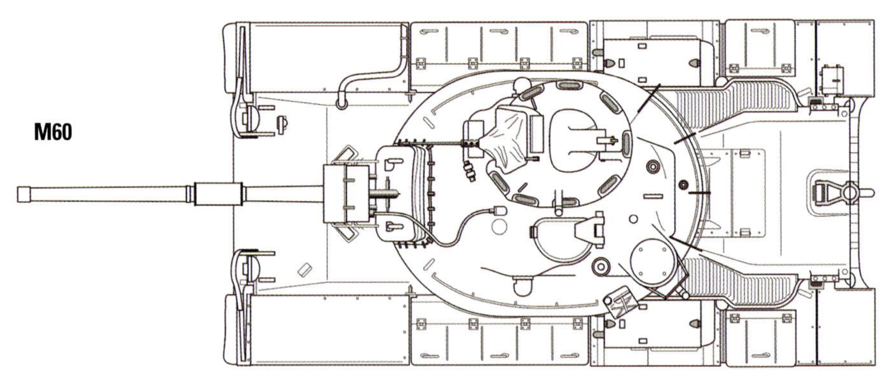

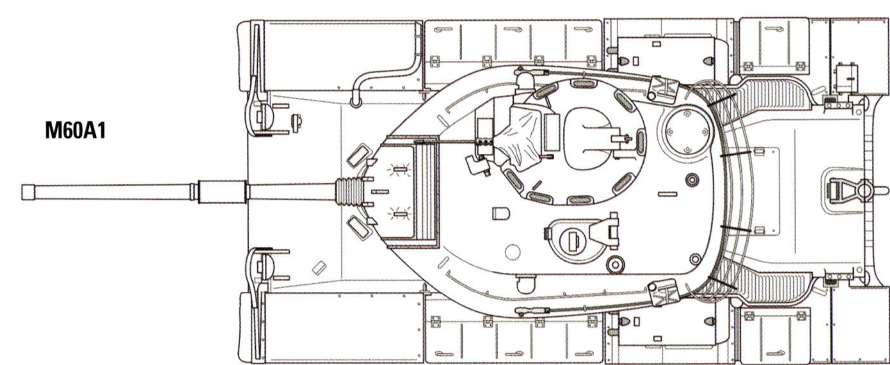

The initial M60 production was equipped with an oval, almost round turret derived from that used on the M48. The M60A1, known during the development phase as the M60E1, used a much-larger, elliptically shaped turret, derived from that of the experimental T95E7 tank. In addition to providing more room for the crew, the new turret included space for eight additional rounds of 105 mm ammunition stowage in the turret bustle. Even more significantly, the larger turret had heavier armor, thus providing additional protection. Not only was the thickness of the gun shield increased from 4.5 inches to 5 inches, the protection of the face of the turret was increased from the equivalent of 7 inches to 10 inches. Similarly, side armor protection went from 3 inches to 5.5-inch equivalency. The 105 mm gun M68 was installed in an M116 mount on the M60 tank, while the M60A1 used an M140 mount. The later M60A3 used an M68E1 gun with thermal shield with the same turret and mount as the M60A1.

Although the M60 was in full production, work continued toward producing an improved tank. While the M60 was equipped with what was essentially an improved M48 turret, work had continued on the larger turret originally proposed for the aborted T95E7 tank. Chrysler was directed on March 21, 1960, to create three pilot tanks with the larger turrets mounted on M60 chassis.

The first of these three pilots, which were given the designation M60E1, was completed on May 6, 1961, and subsequently shipped to Aberdeen Proving Ground. The second pilot, completed twenty days later, went to Detroit Arsenal, while the third, finished on June 30, was sent to Fort Knox.

While the principal difference between the M60 and M60A1 was the larger turret, other improvements, born from the initial experiences with the M60, were incorporated as well. These included improved driver's controls and seats, and slightly improved suspension.

On October 22, 1961, the M60E1 was classified Standard A as the "105 mm Gun Full Tracked Combat Tank M60A1" by OCM 37933. The same action reclassified the M60 as Standard B upon delivery of the new model. Production M60A1 vehicles featured heavier hull armor than that of the M60E1 and, from August 1962, were powered by the Continental AVDS-1790-2A engine, which boasted greater fuel economy and reduced smoke over its predecessor.

The M60A1 would remain in production for twenty years, the last example leaving the Chrysler assembly line in May 1980.

The M60E1 pilot 1, registration number 9B3487, undergoes preliminary testing at the Detroit Arsenal on May 19, 1961. The M60E1 featured an M60 chassis with a different, longer turret, of the type used on the experimental 90 mm T95E7 gun tank. This turret provided better frontal protection and offered more interior space. Further, it lacked the turtleback contours of the M60 turret. Initially, the front and the rear suspension arms were equipped with friction snubbers, but later in the test program these were replaced by hydraulic shock absorbers. *Patton Museum*

The third M60E1 pilot, 9B3486, was completed on June 20, 1961; by July 20 it had arrived at Fort Knox for testing. Painted in white on the side of the turret is "M60E1" over "PILOT 3." "TEST OPERATION" and a yellow circle with a "53" bridge classification are on the glacis. *TACOM LCMC History Office*

The second pilot M60E1, 9B3487, drives over a concrete obstacle during testing. The air cleaner is the late-type side-loading model, lacking the distinctive access-door latch of the earlier version. Small, white recognition stars are on the rear mudguards. *TACOM LCMC History Office*

This M60E1, registration number 9B5605, is serving as a test vehicle at Aberdeen Proving Ground, Maryland, on April 23, 1973. The M60E1 design was standardized, with some improvements, as the 105 mm M60A1 gun tank. It featured the elongated turret of the M60E1. A difference that was not noticeable to the eye was that on the production M60A1s, some areas of the armor were thicker than on the M60 and M60E1. Also, the M60A1 had shock absorbers on the two forward suspension arms, in addition to a shock absorber on the rear suspension arm. *National Archives*

The same M60E1 test vehicle seen in the preceding photo, registration number 9B5605, is viewed from the left side at Aberdeen Proving Ground on April 23, 1973. On the turret, between the rear of the handrail and the turret basket, is a holder for a 5-gallon liquid container. *National Archives*

M60E1 9B5605 is seen from the right side at Aberdeen on April 23, 1973. The elbow-shaped fitting on the rear of the turret roof was part of a holder for a xenon searchlight, on which the searchlight would be mounted during travel or when not in use. *National Archives*

This elevated view of M60E1, registration number 9B5605, on April 23, 1973, reveals details of the turret roof, the glacis, the fenders, and the hull roof. The driver's hatch is popped open. The driver had an M24 infrared periscope on his hatch and three M27 periscopes to the front of the hatch. *National Archives*

The M60A1 with registration number 9B5593 rests on a field between tests at Fort Knox, Kentucky. A flexible, accordion-type dustcover was usually installed over the base of the main gun barrel, and it worked in conjunction with the mantlet cover to seal out dust and fumes from the turret. *National Archives*

An M60A1 main battle tank appears in new condition, with an overall camouflage of semigloss Olive Drab paint. "US ARMY" and the registration number, 09A11071, are painted in white on the toolbox on the fender. The mantlet cover is Olive Drab, while the mud flaps are black. *TACOM LCMC History Office*

A welder attaches brackets to an M60A1 turret interior at the Detroit Arsenal in 1975. The first 360 M60s were made at Chrysler's Newark, Delaware, plant before production of M60s and subsequent models was shifted to the Detroit Arsenal.

A completed turret, with its white-painted basket below it, is married to the chassis of an M60A1 at the Detroit Arsenal Tank Plant, Detroit, Michigan, in 1975. At that time, the plant was completing forty tanks per month; the Army hoped to increase that number to 103.

All M60-series tanks had an 85-inch turret ring diameter. On the M60, sixteen of the fifty-seven 105 mm rounds were stowed as ready rounds, while on the M60A1 and A3, thirteen of the sixty-three rounds were ready service rounds.

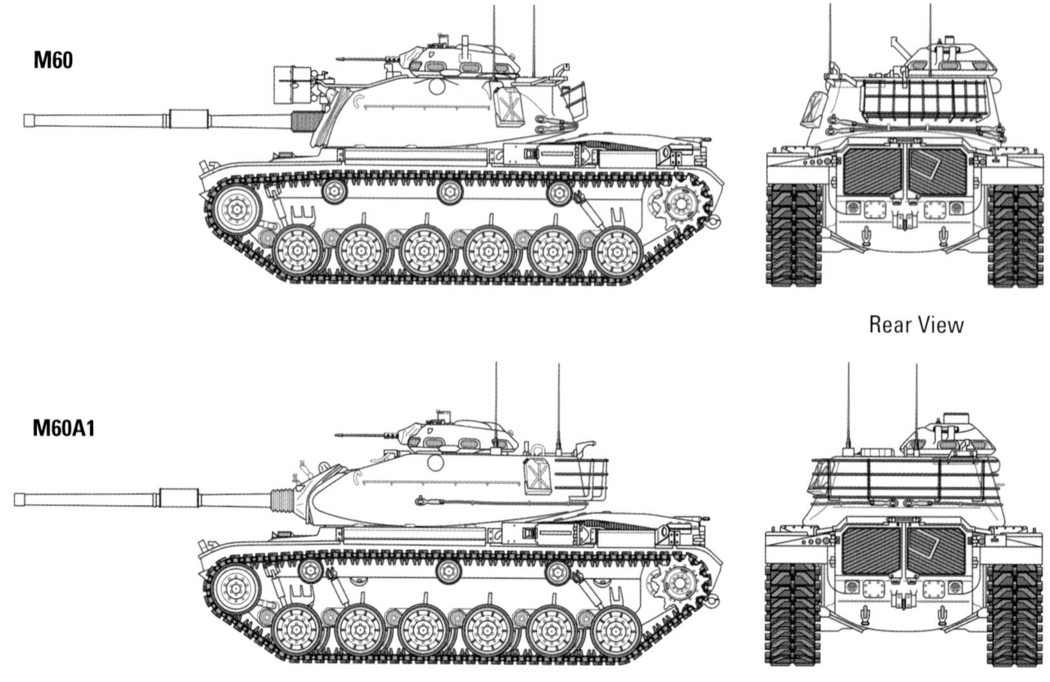

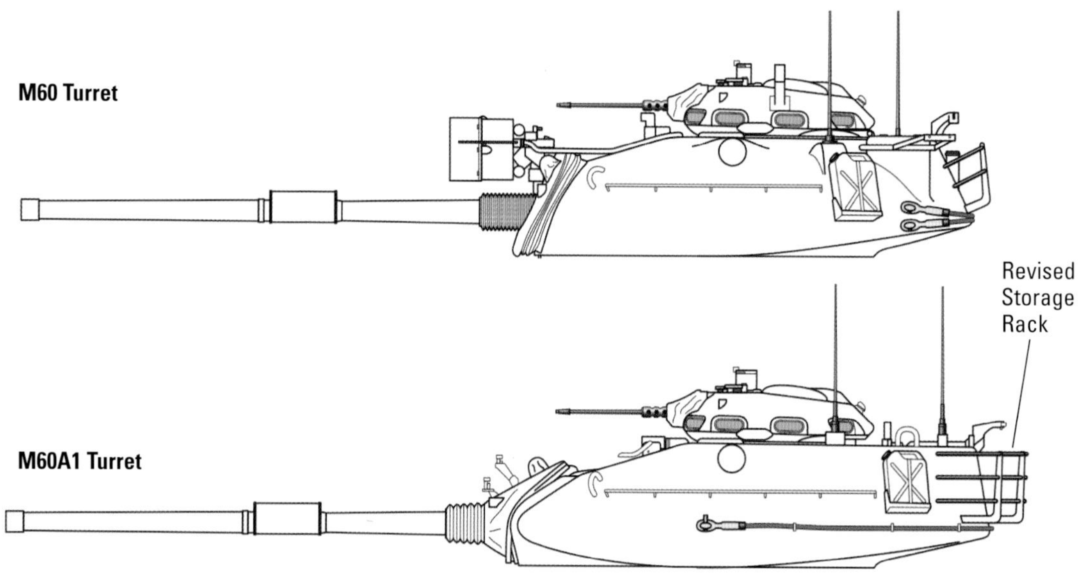

The M60 turret had a 24-degrees-per-second traverse rate, and an elevation range of +10 to –9 degrees. The M60A1 traverse rate was 22.5 degrees per second, and elevation limits of +20 to –10 degrees.

In 1971, a top-loading nonarmored air cleaner was devised for the M60A1. It reduced the amount of dust intake, resulting in longer engine life. Then, in 1972, an add-on stabilizer kit became available for the main guns of the M60A1s, leading to improved accuracy when firing on the move. Finally, a new model of track, the T142, became available. It was a double-pin, steel design with replaceable rubber pads. When M60A1s received these three improvements, they were designated M60A1 (AOS), an example of which is shown here in a vehicle park, with its turret traversed to the rear. *TACOM LCMC History Office*

Following the M60A1 (AOS), further improvements were made to the line. In 1975, the engine underwent a revamping known officially as Reliability Improved Selected Equipment (RISE), and tanks equipped with the revamped engine were designated M60A1 (RISE). Further improvements included the new AN/VSS-3A xenon searchlight, a thermal sleeve for the main gun barrel (absent from these vehicles) to protect it from distortion due to fluctuations in temperature, improved passive periscopes for the commander and gunner, and a night-vision device for the driver. Tanks with these upgrades, such as the one shown here, were designated M60A1 (RISE) (Passive). Brackets for M239 smoke grenade launchers are also visible on the turret sides. *TACOM LCMC History Office*

Three-color NATO camouflage is present on this M60A1. An AN/VSS-1 xenon searchlight is installed above the gun shield, and a dustcover is over the lens. *Chris Hughes*

The M60A1 is observed from the left side. The travel lock for the 105 mm gun barrel is standing up on the rear of the engine deck. Top-loading air cleaners are present on this vehicle. The bogie wheels are the steel type. *Chris Hughes*

An M60A1 is parked next to a German Leopard 1 main battle tank of similar vintage. The headlight assemblies on the M60A1 have been removed from the bottom of the glacis, and plugs have been inserted in their sockets. The bogie wheels are forged aluminum. *Chris Hughes*

On the side of the turret is a British-designed M239 launcher for six smoke grenades. To the front of the 5-gallon liquid container on the turret is a storage box for smoke grenades. *Chris Hughes*

Features on the bottom of the xenon searchlight are in view, including a small, rectangular vent with a screen over it. The dustcover for the gun shield incorporates two extensions to cover the lifting eyes on the shield; the right one is in view. *Chris Hughes*

Openings in the dustcover of the gun shield are for the gunner's sight (the larger opening on the right-hand side) and for the coaxial machine gun on the opposite side. The sides of the dustcover incorporated an accordion design, to allow the cover to flex with the elevating and depressing of the guns. *Chris Hughes*

By the time the US Marine Corps deployed its M60A1 tanks to the Gulf War in late 1990 and early 1991, the vehicles were largely equipped with explosive reactive armor (ERA). This consisted of tiles made of an explosive sandwiched between two steel plates; the tiles were arrayed at a short distance from vulnerable areas of the forward hull and the turret. A hit from an antitank shell or missile detonated the tiles, creating a force in the opposite direction, thus defeating or greatly diminishing the projectile's ability to penetrate the tank. These photos depict an M60A1 equipped with ERA. *Chris Hughes*

The M60A1 with ERA is painted in one of the sand camouflage colors, evidently Desert Tan, used in Operation Desert Storm in 1991. The inverted V marked in black on the tiles on the side of the turret was the recognition sign for Coalition forces in that war. *Chris Hughes*

A comparison of this photo with the preceding one shows that the pattern of the reactive-armor tiles on the left side of the turret was not identical to that on the right side. Spaces were left between the tiles for the rangefinder domes and the M239 smoke-grenade launchers. *Chris Hughes*

The layout of the ERA tiles on the front of the turret and on the bow of the M60A1 is illustrated. A tier of the tiles is on the upper part of the lower front plate of the bow. Gaps in the tiles on the front of the turret provide clearance for the gunner's sight and the coaxial machine gun. *Chris Hughes*

The ERA tiles are fastened with hex screws and washers to brackets that are attached to the turret and the hull. Four fasteners per tile are used. Tiles on the right front of the turret are shown. *Chris Hughes*

Several of the brackets for ERA tiles are viewed from below on the right front of the M60A1 turret. *Chris Hughes*

The interior of the turret of an M60A1 is observed from behind the breech of the 105 mm M68 gun in an M140 mount. On the left side of the breech is the operating handle for the breechblock. To the left are 105 mm ready rounds. To the right is the gunner's station, including his seat and hand controls. *Author*

This is a vehicle-commander's-eye view down into the turret from inside the cupola. A fire extinguisher is to the rear of the ready rounds to the left. The gunner's seat is to the right of the 105 mm gun breech. Toward the upper right is the commander's control handle, below which is the turret traverse gearbox assembly. The curved objects on the floor below the commander's station are boxes for .50-caliber ammunition (the two inner boxes) and 7.62 mm ammo (the outer boxes). *Author*

Radio equipment is installed in the turret bustle of the M60A1. The view is toward the right rear of the turret. To the far left is the commander's seat, with its bottom in its folded position. *Author*

In a view of the left side of the turret bustle, to the left is a guard for ejected 105 mm casings, and to the right are tubes for storing 105 mm rounds. Eighteen rounds were stored in this rack. *Author*

The driver's compartment of an M60A1 is viewed facing forward from above the platform of the turret basket, with several 105 mm ready rounds to the left. For this model of the tank, the steering wheel was replaced by a steering bar, seen here below the periscopes toward the top. At the center are the instrument panel and the master control panel, to the left of which is the seat and its support. *Author*

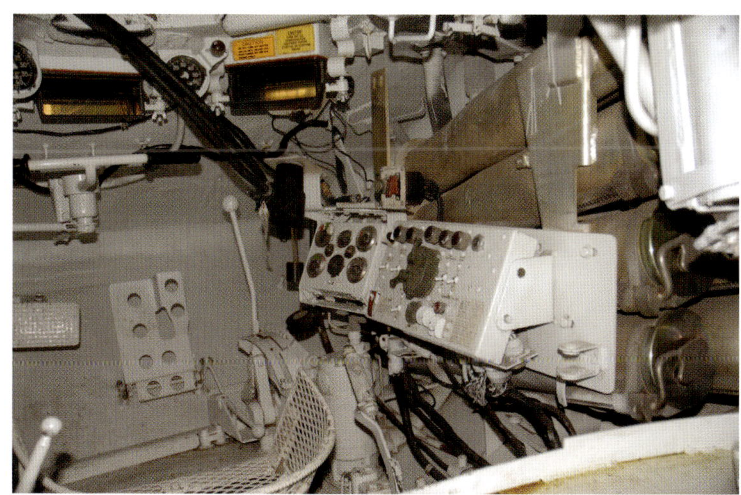

The instrument panel and master control panel are viewed from a closer perspective. To the right are tubes for storing 105 mm ammunition. Below the steering bar are the brake pedal, *left*, and the accelerator pedal, to the right of which is the transmission shift lever. *Author*

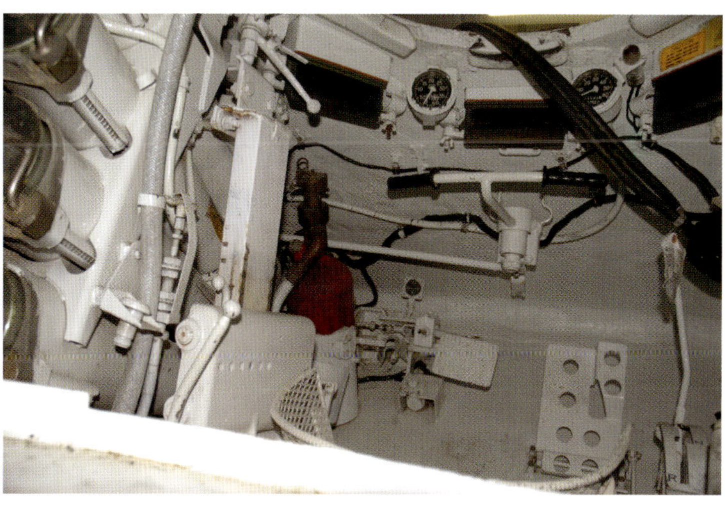

In another view of the driver's compartment of the M60A1 from the rear, at the center is a fixed fire extinguisher; another one is next to it but is obscured from view. To the left of the extinguisher is the support post for the driver's seat. To the left are storage tubes for 105 mm ammunition. The gauges at the top are the tachometer / hour meter (*left*) and the speedometer/odometer (*right*). *Author*

CHAPTER 3

M60A2 "Starship"

The M60A1E1 grew out of US Army planning in the late 1950s for eventually arming its tanks with guided missiles, which the Army saw as the wave of the future in ground firepower. The turret of the M60A1E1 was distinctly different from those of the M60 and the M60A1, designed around an XM81 Shillelagh Combat Vehicle Weapon System (CVWS) gun launcher, which could fire a conventional projectile or an XM13 Shillelagh guided missile. *TACOM LCMC History Office*

The M60A2, often informally dubbed the "Starship," was the result of the prevailing belief in the late 1950s and early 1960s that future generations of tanks would be armed with missiles rather than guns. The initial premise was that this would be a ready adaptation of a fleet of existing tanks to utilize the armament of a future generation of tanks—namely, the XM81 Combat Vehicle Weapons System (CVWS) Shillelagh.

The CVWS would fire either the XM13 Shillelagh missile or a 152 mm conventional round, the latter equipped with a fully combustible cartridge—that is, there is no brass to dispose of or recycle after firing. The CVWS featured a gun/launcher about half the size and weight of the 105 mm gun of the M60A1. Once standardized, this gun launcher was the XM162E1. The M551 Sheridan would be equipped with a similar CVWS.

Four possible turrets, types A through D, were proposed. Types C and D derived from conventional tank turrets, while A and B were special, compact turrets. The trial vehicles were designated M60A1E1, and turret type B was chosen for further development.

Procurement of 243 turrets to be retrofitted to existing chassis was approved for fiscal 1966 funding. A further 300 tanks were to be produced with fiscal 1967 funds. The Shillelagh-armed vehicles were designated M60A2. Chrysler records indicate that 526 conversions were preformed from 1973 to 1975, while Army records reflect 540 during the same time. The new vehicles began to be issued in 1974, arriving in Europe in 1975.

Despite the effort, the end result was disappointing, and the Shillelagh was removed from service in 1981. At that time, most of the M60A2 chassis were rebuilt as either M60A3 gun tanks or AVLB.

The M60A1E1 turret had a narrow central section to house the XM81 CVWS, presenting a smaller frontal target. The vehicle commander occupied a cupola armed with a .50-caliber machine gun. The gunner and the loader were seated under the low parts of the turret to the sides. *TACOM LCMC History Office*

The second M60A1E1 pilot, registration number 9B3140, is viewed from above during evaluations at Fort Knox, Kentucky, on July 27, 1966. The gunner's hatch and the head of his M50 periscope are on the right side of the turret. The commander had an M51 periscope. *TACOM LCMC History Office*

The "XM81E2 Cannon, Gun-Launcher, 152 mm" was the predecessor of the XM81E3 used in the M60A1E1, which was subsequently standardized as the M162 and used in the M6A2. *Rock Island Arsenal Museum*

This tank had Army registration number 9B3129 and was one of two M60A1E1s, the pilots for the M60A2. The XM81 Shillelagh CVWS was standardized as the M162 152 mm gun launcher. *TACOM LCMC History Office*

M60A1E1 9B3120 was photographed during evaluations at Aberdeen Proving Ground, Maryland, on February 10, 1966. A hinged boarding ladder rests on the left side of the glacis; the purpose of the bow-shaped assembly hinged to the top of the glacis is not clear. *National Archives*

This early M60A1E1, 9B4470, had the proposed "A" turret (turret style "B" was used on production vehicles). The vehicle was photographed outside the Detroit Arsenal. *TACOM LCMC History Office*

Another view of 9B4470, which illustrates the excellent visibility that the commander had through eleven vision blocks around the base of the cupola. An AN/VSS-1(V) xenon searchlight was on the left side of the turret. *FCA North American archives*

The Army ordered 243 Shillelagh-armed production turrets to be installed on M60 chassis; initially these vehicles would be designated M60A1E1. This one, 9B4470, had a four-tube smoke-grenade launcher on the turret above the gunner's periscope and a hinged ladder on the glacis. *National Archives*

Tests are underway on two vehicles that shared the XM81 Shillelagh CVWS, at White Sands Missile Range, New Mexico, on October 8, 1965. On the left is one of the M60A1E1s, and on the right is a Sheridan armored-reconnaissance/airborne-assault vehicle (AR/AAV). *National Archives*

US Army registration number 9B4057 was the M60A1E1 advanced production engineering prototype, seen on December 21, 1967. "TANK OPERATION" is stenciled in white on the upper center of the frontal plate of the hull. An extra amber light is on the left brush guard. *TACOM LCMC History Office*

Details of the turret of M60A1E1 9B4057 are visible in an image from December 21, 1967. Baskets are attached to the sides and rear of the turret bustle, with a ventilator located in the center rear of the basket. A travel bracket for the xenon searchlight is above the basket. *TACOM LCMC History Office*

M60A1E1, registration number 9B4057, is viewed from the left side. As was the case with the production M60A2, on top of the mantlet was an armored box containing the infrared transmitter, which communicated with the Shillelagh missile to adjust its course during flight. The height of the M60A1E1 to the top of the cupola periscope was 2 inches more than that for the M60A1. *TACOM LCMC History Office*

In this overhead view of the M60A1E1 advanced production engineering prototype, it can be seen that four-tube smoke-grenade launchers have been mounted on each side of the turret bustle, inside the turret basket. Casting numbers are on the turret roof in front of the cupola. *Patton Museum*

Vehicles with the Shillelagh CVWS turret mounted on the M60A1 chassis initially were designated the M60A1E2 and later standardized as the M60A2. This M60A1E2 was US Army registration number 09A05667. A weatherproof cover is installed over the mantlet/turret joint. *Patton Museum*

A driver's head is visible below the bustle of the turret, which is traversed to the rear, on M60A1E2, US Army registration number JK003L. There were three shock absorbers per side on the vehicle, for the two front bogies and the rear one. The front one is not visible from this angle. *TACOM LCMC History Office*

The M60A1E2 was standardized as the M60A2 152 mm gun tank. A noticeable difference between the earlier vehicles and the production M60A2s was that the latter lacked the bore evacuator on the 152 mm gun. This example was registration number 09A05767. *TACOM LCMC History Office*

An advantage of the M60A2 was that its 152 mm gun was compact and did not protrude past the front of the vehicle, or past the rear when the turret was traversed aft. Protruding from the upper left corner of the front of the mantlet is the muzzle of a 7.62 mm coaxial machine gun. *TACOM LCMC History Office*

This diagram and the following one illustrate the difference between the hull rear of the M60A1E1 / early M60A2 and the late M60A2. The M60A1E1 and early M60A2 had a hull rear similar to that of the M60 and M60A1, with no jutting fairing below the engine compartment door.

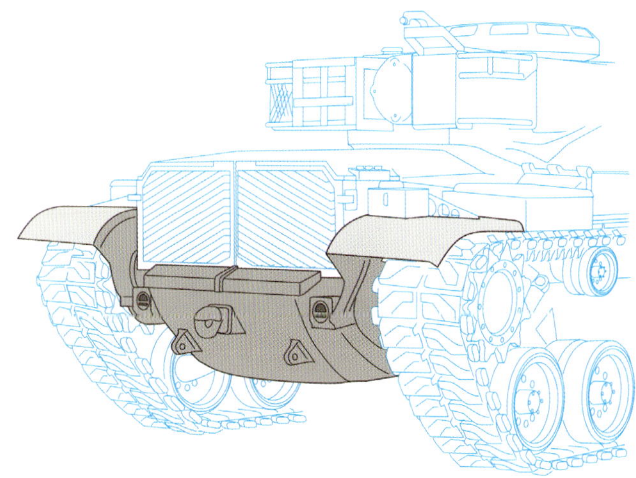

Late in M60A2 production, a new 152 mm gun launcher was introduced, without a bore extractor. In place of the barrel-mounted extractor, a closed-breech-scavenger system was installed, resulting in a fairing on the rear of the hull to house two compressors and two air bottles.

The M37 tank-gunnery trainer was fashioned from an M60A1E1 turret with some panels cut from the cast armor. A walkway was installed around the base of the turret, with support arms stretching up to the sides of the upper part of the turret, and steel mesh guards installed. *Rock Island Arsenal Museum*

An early M60A2 tank is on a test incline at Letterkenny Army Depot in Pennsylvania on April 22, 1975. The 152 mm gun launcher was being tested for its force-margin recoil on this specific slope. A motorcycle helmet, for blast protection, is on the turret roof in front of the cupola. *National Archives*

An M60A2 152 mm gun tank of the Jacques Littlefield collection in a fine state of preservation is parked next to a SS-1b Scud A tracked missile launcher. Despite the impression the vehicle gave of a much-taller profile than the M60, owing to the bulky cupola, the M60A2 was only 4 inches taller. *Chris Hughes*

The M60A2 is viewed from the front, with the turret traversed slightly to the right. A dustcover is over the lens of the xenon searchlight, and the power cable for the searchlight is connected to the receptacle on the turret roof. The searchlight for the M60A2 was the AN/VSS-1, producing 100 million candlepower at 2.2 kilowatts. When the usual black infrared filter of the AN/VSS-1 was replaced with a pink IR filter, the searchlight was designated AN/VSS-2. *Chris Hughes*

On the top of the gun shield is the box-shaped housing for the infrared transmitter, for the Shillelagh missile system. The object atop the baggage rack on the turret bustle is a camouflage net. *Chris Hughes*

The loader's hatch door is open on the left side of the turret. The door is resting against a stop that holds it open at this angle. The air cleaner is the top-loader version. Fine details of the sprocket are in view. *Chris Hughes*

In a frontal view of the M60A2, the housing for the gunner's M50 periscope is to the front of his open hatch door. The narrow profile of the turret is evident. An M85 .50-caliber machine gun was installed in a module on the right side of the cupola. *Chris Hughes*

Details of the tracks and their end connectors, the left air filter and rear storage box, and the left side of the turret are in view. To the far right is the left rear lifting eye. Two small lifting rings are visible on the side of the cupola. Jutting from the upper rear of the baggage rack is a storage bracket for the xenon searchlight. *Chris Hughes*

The left air cleaner housing is viewed close-up. It is a top-loading type, with a triangular lifting ring on the side. *Chris Hughes*

In a left view of the turret, to the left is the dustcover for the Shillelagh mount, at the center is the AN/VSS-1 searchlight with its cover installed, and to the right are the loader's hatch and the cupola. A grab handle is on the side of the searchlight. *Chris Hughes*

Further details of the dustcover for the Shillelagh mount are displayed. At the upper left, the armored door for the infrared transmitter is lowered. The tube at the upper left corner of the gun shield is the port for the coaxial 7.62 mm machine gun. To the left is the barrel of the M85 machine gun, with a flash hider on its muzzle. *Chris Hughes*

Details of the dustcover of the searchlight are shown. To the left is the end of a tow cable, with the eye secured in a bracket on the side of the turret. *Chris Hughes*

A close-up view of the left side of the turret of the M60A2 includes the rear of the xenon searchlight and its mounting bracket; the power cable for the searchlight, which is plugged into the rear of the searchlight and into the receptacle on the turret roof; the loader's hatch, with 152 mm ready rounds visible inside; and the hatch door and the cupola. To the rear of the hatch door are a lifting ring and a plate covering a mounting location for an antenna. On the turret above the hatch door is the trademark of General Steel Castings Corporation. *Chris Hughes*

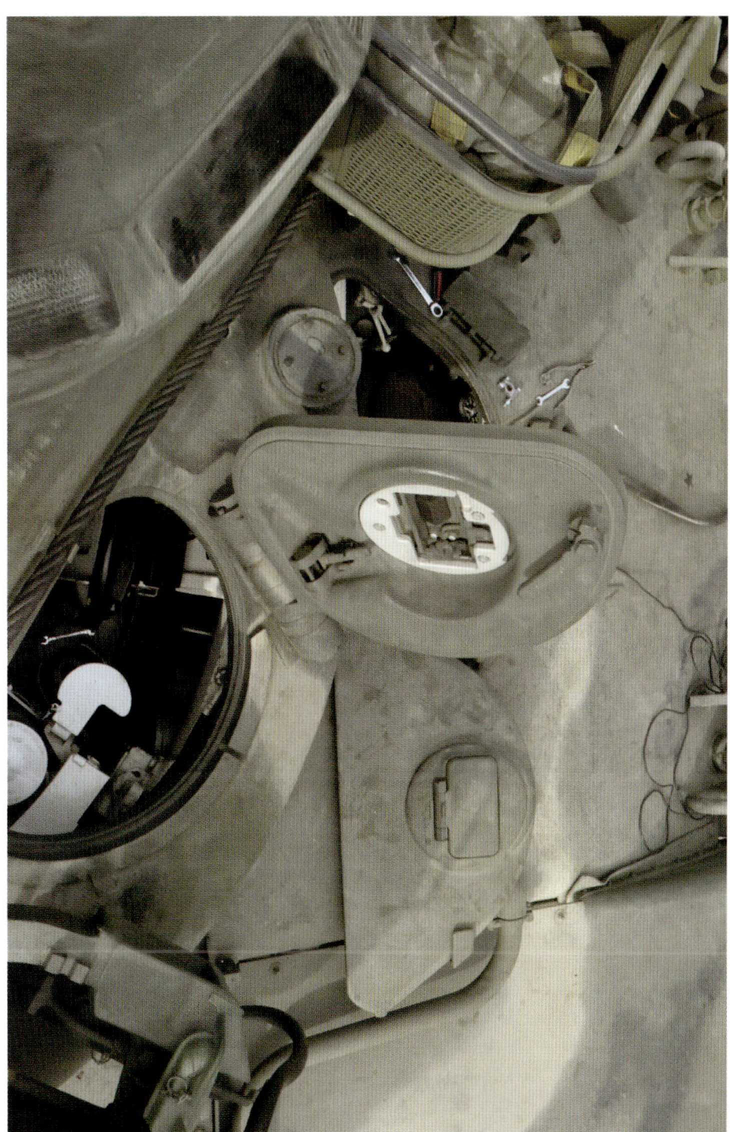

The left side of the turret is viewed from above, with the bow of the tank to the right and the turret traversed right to approximately the four or five o'clock position. To the upper left is the cupola. Below the loader's hatch door are the driver's hatch and its open door. *Chris Hughes*

In a view of the front of the turret of the M60A2 from the upper rear, a tow cable is secured to a bracket on the side of the turret. Also visible are the box-shaped housing for the infrared transmitter, the front lifting ring, and the receptacle for the power cable to the searchlight. *Chris Hughes*

In a front view of the cupola, four vision blocks are in view; at the top is the front of the commander's periscope and its hood. The cover for protecting the periscope glass when not in use is folded up between the hood and the top of the periscope. *Chris Hughes*

The cupola of the M60A2 is viewed from the rear, showing three of the vision blocks, the open hatch door, and, *to the right*, the module for the .50-caliber machine gun. Heaped in the foreground is a camouflage net. *Chris Hughes*

Expanded-steel mesh is on the lower approximately two-thirds of the baggage rack, as seen from the right side. Two spare track shoes are attached to the rack, as is a 5-gallon water container. To the right is the gunner's hatch. *Chris Hughes*

The entire turret is viewed from the right side, with the gunner's and cupola hatches open. The curved pipe in the lower foreground is the exhaust for the personnel heater. *Chris Hughes*

The open hatch to the gunner's compartment is observed from the rear. The tab jutting from the hatch door is a padlock hasp. To the immediate rear of the hood for the gunner's periscope is a grab handle. *Chris Hughes*

CHAPTER 4

M60A3

The M60A3 (an early one is shown in 1975) incorporated numerous improvements implemented during the production of the M60A1, as well as some new twists. These innovations included a thermal sleeve for the 105 mm gun barrel, an AN/VGG-2 laser rangefinder and solid-state ballistic computer, passive night sights for the commander and gunner, and the new, compact AN/VSS-3A searchlight. Further improvements would be installed during M60A3 production, such as a tank thermal sight (TTS) for the gunner. *National Archives*

Although the M60A1 had seen many upgrades, including the top-loading air cleaner, the add-on stabilization (introduced in 1972), and the Reliability Improved Selected Equipment (RISE) package in 1975, with its AVDS-1790-2C engine, there was still room for further improvement.

In 1978, the British-designed M239 smoke grenade launcher, the M240 coaxial machine gun, the ruby laser rangefinder, and the M21 ballistic computer and thermal shroud for the main gun were added, and the resultant improved vehicle was given a new designation: M60A3. Not only did the M60A3 replace the M60A1 tanks coming off the production line, but older tanks, both M60A1 and M60A2, were also modified to M60A3 standards.

Additionally, many of the M60A3s were equipped with the AN/VSG-2 sighting system, becoming known as the M60A3 tank thermal sight (M60A3 TTS). Utilization of the T142 model track with replaceable rubber shoes reduced operational cost.

The US Army began receiving the newer M1 Abrams tanks in the 1980s, but for a time the US Marines stayed with the M60A1, of which they had purchased 578 new examples. Later, when the Army acquired sufficient excess and surplus Abrams tanks, these became available to the Marine Corps, allowing the USMC to transition to the M1 at extremely reduced cost.

Although the M60A3 was withdrawn from US Army service in 2005, it remains to this day a frontline vehicle for many nations.

The same M60A3 in the preceding photo, registration number 09A03970, is on view again. Many of the new systems on the M60A3 had been tested on the product-improved M60A1 and M60A1E3 programs, including the thermal sleeve and the T142 tracks with octagonal pads. *TACOM LCMC History Office*

M60A3 09A03970 undergoes evaluation on a test track. The short pole behind the two radio antennas was a crosswind detector, to provide wind data to the fire-control computer. The old AN/VSS-1(V) searchlight had been replaced by the more compact AN/VSS-3A type. *TACOM LCMC History Office*

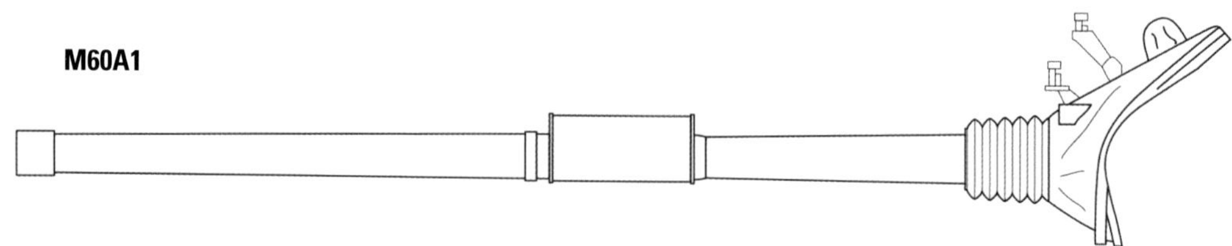

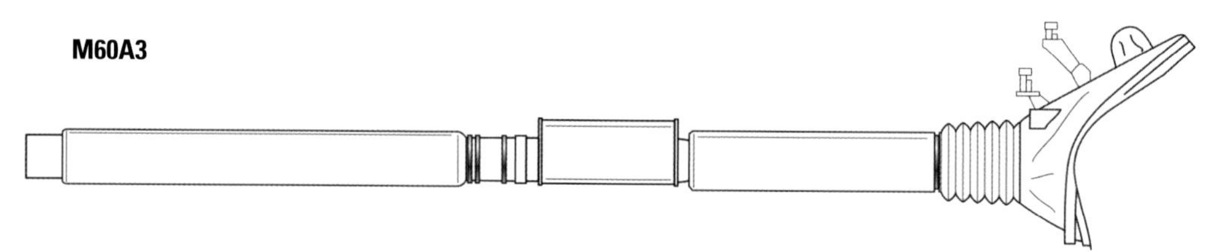

The M60A3 had a thermal sleeve on the 105 mm gun barrel, with one section to the front of the fume extractor and one to the rear of it. The thermal sleeve was designed to prevent factors such as sunlight and changes in ambient temperature from distorting the gun barrel, which would affect accuracy.

An M60A3 exhibits several of the distinctive features of that model: a thermal sleeve over the 105 mm gun, an M239 smoke grenade launcher and a smoke grenade storage box on the side of the turret, and a redesigned right dome for the new laser rangefinder. *TACOM LCMC History Office*

A disruptive camouflage scheme, evidently of Olive Drab, tan, black, and medium gray, has been applied to this M60A3. The thermal sleeve was designed to prevent distortions to the gun barrel from thermal expansion and contraction, which affects the accuracy of the weapon. *TACOM LCMC History Office*

An M60A3 painted in three-color NATO camouflage lacks dustcovers on the main and cupola gun mounts, enabling a view of the respective gun shields of those mounts. The M239 smoke grenade launchers are not installed, allowing a full view of their mounting brackets. *Chris Hughes*

Although the laser rangefinder on the M60A3 used the dome on the right side only, the left dome was retained, possibly because it was more expedient to retain it than to plug the hole that would result from removing it. *Chris Hughes*

The tracks on this M60A3 are the T142 type. These featured two replaceable, octagonal rubber pads on each shoe. On the upper part of the 105 mm gun shield are mounting brackets for a searchlight. *Chris Hughes*

All three of the driver's periscopes are raised on this M60A3 on display at the Central Coast Veterans Museum, San Luis Obispo, California. *Chris Hughes*

An M60A3 with ERA installed is viewed from the right front. The tracks are T142s. Because the reactive armor was over the places on each side of the turret where the smoke grenade storage boxes otherwise would have been mounted, these boxes were mounted on the sides of the air cleaner housings, as seen here. *Chris Hughes*

As seen in a frontal view, the ERA tiles to the sides of the driver's hatch are not symmetrical; those on the right side are mounted above the right fender, while those on the opposite side are mounted atop the hull. The tile normally immediately above the left tow eye is missing; its counterpart above the right tow eye is present. *Chris Hughes*

The arrangement of ERA tiles on the right side of the turret is illustrated. The laser rangefinder dome on the right side of the turret was provided with an armored door, to protect the sensitive electronic equipment inside. *Chris Hughes*

The ERA array on the left side of the turret is shown. A box for smoke grenades is mounted on the side of the air cleaner housing. *Chris Hughes*

A grab handle is mounted on the ERA tiles on the left side of the M60A3 turret, secured with two of the hex screws for securing the tiles. The smoke-grenade storage box has a hinged lid on top, with two latches. *Chris Hughes*

The upper-right corner of the right transmission-access grille-door is to the lower left in this view of the turret bustle and the baggage rack. Above the left side of the bustle is the crosswind sensor, a new addition with the M60A3. This device fed data on crosswind to the ballistic computer, to correct for the effect of wind on the trajectory of shells. *Chris Hughes*

The array of nine ERA tiles on the right side of the turret bustle of the M60A3 is viewed from below. The three rear tiles are wider than the others. Also in view are the tiles on the side of the turret, showing the supports for the brackets, which are welded to the turret. At the bottom is the right storage box for smoke grenades. *Author*

The bogie wheels on this M60A3 are the steel type. Variations in the sizes of the ERA tiles are noticeable. The three hinges on the rear of the door on top of the air cleaner housing are visible. *Author*

The tiles on the 105 mm gun shield are arranged so that there is no impediment to the gunner's view through his periscope. The elbow pipe on top of the right side of the driver's compartment is the bilge-pump outlet. *Author*

Installing ERA tiles on the M60A3 required a complicated and varied assortment of clips, brackets, and posts. Some of the ERA tiles and their mounting brackets on the gun shield are illustrated. *Author*

In a view of the driver's compartment of an M60A3, at the top is the hatch, below which are three crash pads (the periscopes are not installed), and the tachometer / hour meter (*left*) and the speedometer/odometer (*right*). Farther below are the steering bar, two fixed fire extinguishers, the brake and accelerator pedals, and the transmission gearshift lever. *Chris Hughes*

In the lower front of the driver's compartment are, *left to right*, the turret-seal pump, the headlight-dimmer foot switch, the brake pedal, the accelerator pedal, and the transmission gearshift lever. *Chris Hughes*

On the right side of the driver's compartment of the M60A3 are the instrument panel (*front*) and the master control panel. Below the latter is a bin for storing rations. *Chris Hughes*

The driver's instrument panel is viewed close-up. Immediately above it are the dust-detector warning light (*left*) and the smoke-generator switch (*right*). That switch controlled the vehicle-engine exhaust smoke system (VEESS), which was new for the M60A3. This system injected diesel fuel into the exhaust system to produce a thick smoke screen. *Chris Hughes*

Facing the right front of the interior of the cupola, to the left are the daylight body and diopter of the commander's M36E1 periscope, to the right of which are the infrared body and diopter for night vision for the commander's periscope. The black cylinder below the IR diopter is the handle for the cupola traverse control. *Chris Hughes*

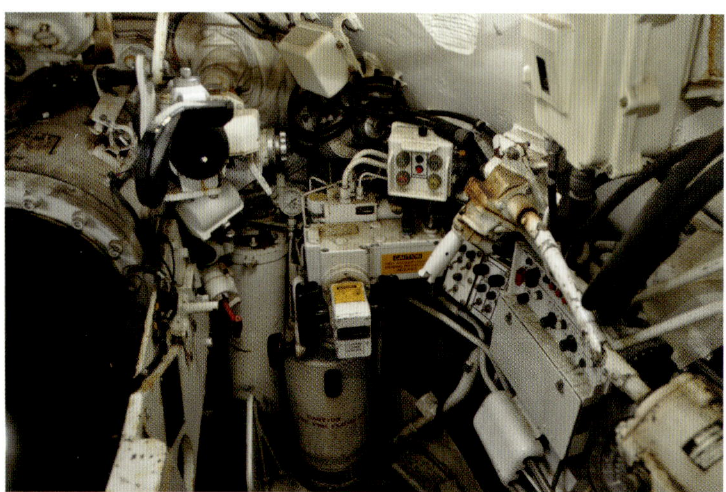

In a gunner's view of his station in an M60A3, toward the left, alongside the 105 mm gun, is the sight, M105D telescope, to the lower right of which is the gunner's power control, which includes two black handgrips. To the upper right of the power control is the manual traversing handle, below which are switch boxes. *Chris Hughes*

The gunner's compartment is viewed from the vehicle commander's perspective, with the gunner's seat at the bottom and the power control above it. The dial to the right of center is the azimuth indicator. *Chris Hughes*

CHAPTER 5

M60 Derivatives

Armies maneuvering in the field require a wide range of support vehicles, including specialized engineering vehicles. These are particularly important when it comes to overcoming streams and gorges, and when breeching heavy fortifications. To ease the logistics burden of supporting these specialized vehicles, they often share chassis and components with the organizations' primary vehicles.

In the case of the M60 series, the two primary engineering vehicles based on the chassis were the M60 AVLB (armored-vehicle-launched bridge) and the M728 CEV (combat engineer vehicle).

Production of AVLBs based on the M60A1 chassis began in 1964 and continued until 1973, with Chrysler turning out 373 examples. In 1978, an additional fourteen units were constructed, with a final six being built in 1980–81.

Two lengths of bridge were produced, described as 40 foot and 60 foot, although the actual lengths were 43 feet and 63 feet, respectively. The scissor-type bridges, which were made of aluminum, were launched and retrieved hydraulically, and the AVLB crew could remain in the vehicle through the entire operation.

The chassis differed from that of a combat tank in that the turret was removed, and the opening and the driver's position were relocated from the front of the tank to a position on the left side of the area below the turret ring. The vehicle commander was positioned to his right, and both stations were equipped with a vision cupola.

In October 1963, authorization was given to divert three M60A1 tanks from production to be converted into T118E1 combat engineer vehicles. These three prototypes were delivered in March 1964, and one each underwent evaluation at Aberdeen Proving Ground, the Tank Automotive Command, and Fort Knox. In November 1965, the vehicle was classified as Standard A and designated the M728 combat engineer vehicle. The vehicle was armed with a 165 mm M135 short-barreled gun, which fired demolition charges and was equipped with a dozer blade and a lifting A-frame. The turret of the vehicle was a modified M60A1 turret. Chrysler produced 243 of the vehicles from 1966 through 1972. This was later augmented by the RISE-equipped M728A1, produced at Anniston Army Depot, utilizing new-build turrets produced by Chrysler, which were installed on former M60A2 hulls rebuilt at Anniston.

The concept of armored-vehicle-launched bridges (AVLB) dated back to the 1940s. They entailed a tracked, armored vehicle with the means of transporting and laying a scissors bridge under combat conditions. From 1964 to 1973, Chrysler manufactured 373 AVLBs based on M60A1 chassis, with twenty more produced up to 1981. They carried a 43- or 63-foot aluminum folding bridge with a capacity of 63 tons. An M60A1 AVLB is here deploying its bridge at Fort Hood, Texas, in 1975. *National Archives*

Based on the M60A1 chassis, the M728 combat engineer vehicle was a workhorse CEV equipped with a bulldozer blade, an A-frame boom attached to the turret, and a winch. For demolition purposes, it was armed with a 165 mm M135 short-barreled gun, with a supply of thirty rounds of high-explosive, plastic (HEP) ammunition. Typical uses of the vehicle were blasting or bulldozing obstacles, recovering other vehicles, and general hoisting/lifting operations. Shown here in the field at Fort Hood, Texas, in 1975, is the M728 with registration number 9B9281. *National Archives*

As seen in a photo of M60 AVLB, registration number 9B3057, for transport, the scissors bridge was folded, and it rested on the bridge seat, above the rear of the engine deck. At the front of the vehicle were a boom, outrigger, and tongue, along with big hydraulic cylinders. *Military History Institute*

The crew of M60A1 AVLB 9B3057 emplaces the scissors bridge. As the hydraulic cylinders extended the bridge, the boom and outrigger assemblies supported the rear of the bridge; the outrigger is the structure resting on the ground to the front of the tank. *Military History Institute*

Poised here in an upright position is a scissors bridge being deployed by an M60A1 on October 27, 1967. The outrigger may be seen on the ground below the bridge. Fixed-length cables in the scissors bridge acted to pull it open when it was being lowered into position. The bridge could be retrieved from either end. *TACOM LCMC History Office*

On the hull deck of the AVLB where the M60A1 turret had been removed were cupolas for the vehicle commander and the operator/driver, who composed the two-man crew of the vehicle.

M60A1 AVLB, registration number 9B3057, is seen from the right front with the scissors bridge in the travel position. The bridge had a roadway width of 12.5 feet. It took between two and five minutes to emplace the bridge, and about ten minutes to retrieve it, with armor protection. *TACOM LCMC History Office*

The same M60A1 AVLB shown in the preceding photos is in the process of laying a scissors bridge. The crossbeam with the two pointed projections on top of it, lying across the rear part of the engine deck, is the bridge seat, which supported the scissors bridge in the travel position. *TACOM LCMC History Office*

M60A1 AVLB, registration number 9B3057, has just laid a scissors bridge. The piston rod of the tongue cylinder is visible between the front of the chassis and the rear of the bridge. Above the glacis, visible here to the front of the operator's cupola, is another large cylinder, called the overhead cylinder. The M60A1 AVLB weighed a total of 61 tons when the 60-foot scissors bridge was emplaced on it. *TACOM LCMC History Office*

An M728 CEV, registration number 9B7113, rests on a riverbank. The A-frame boom was mounted on, and pivoted on, tubular supports that jutted from both sides of the turret. The headlights and brush guards were elevated, to clear the bulldozer blade in the travel position. *Rock Island Arsenal Museum*

M728, CEV registration number 9B7112, sits on a Tractporter semitrailer transporter. Although not highly visible from this angle, a winch with a 25,000-pound pulling capacity was mounted on the center rear of the turret. The A-frame boom could hoist 17,500 pounds with a single line. *TACOM LCMC History Office*

The same M728, 9B7112, is seen from the front left while loaded on a transporter. The cupola lacks the .50-caliber machine gun, and light-colored tape has been placed over two of the vision blocks. Jutting at an angle above the left side of the turret is the holder for the stay line. *TACOM LCMC History Office*

A good idea of the articulation of the road wheels and the track may be gleaned from this photograph of the M728 with CEV registration 9B7112 disembarking from a semitrailer. The raised bulldozer blade (or moldboard, in Army tech-manual terminology) is almost touching the ground. *TACOM LCMC History Office*

The lifting A-frame of this M728 has been raised. During heavy-lifting operations, the dozer blade could be lowered to provide stability. This vehicle is painted in the four-color MERDC camouflage scheme. *US Army*

CHAPTER 6

Field Use

An M60 is put through its paces at the Armor Board, Fort Knox, Kentucky, in early 1960. This was part of the final test phase for the tank before it was placed into full production. "M60" is stenciled in white on the turret. *National Archives*

Trainees check out in M60, registration number 9B3181, at Area 72 at the US Army Training Center, Armor (USATCA), at Fort Knox, Kentucky, in April 1961. This is an early-production M60 and lacks snubbers or shock absorbers. Those devices would be added later in production. *National Archives*

M60, registration number 9B3181, has just forded a river at Fort Knox on August 5, 1961. The vehicle was outfitted with an improvised underwater-fording kit, including a conning tower with bent-rod steps welded to it, two tall exhaust stacks, and various waterproof seals and covers. *National Archives*

M60 9B3713 at Fort Benning, Georgia, on November 15, 1961, has "CROOK" painted above the "35" on the turret. This was one of the first 300 M60s, which were delivered without .50-caliber M85 machine guns in the cupola, having instead a machine gun mounted externally on the cupola. *National Archives*

The same M60 with "CROOK" marked on the turret advances through a smoke screen at Fort Benning, Georgia, on November 15, 1961. Although the exterior machine gun mount on the cupola was intended for a .50-caliber machine gun, this one is armed with a .30-caliber machine gun. *National Archives*

An AN/VSS-1(V) xenon searchlight is installed on an M60 tank at Aberdeen Proving Ground, Maryland, on March 25, 1963. This searchlight provided both white-light and infrared illumination. The power cable for the searchlight ran from its rear to a receptacle on the turret roof. *National Archives*

The crew of M60 9B5240, from E Troop, Second Squadron, 3rd Armored Division, pause in a field near Weikershof, West Germany, while scouting during a field problem in October 1963. This early M60 has an M2 .50-caliber machine gun on the external cupola mount. *National Archives*

Troops of Company C, 1st Battalion, 32nd Armor Regiment, assume a position in a treeline during Operation Big Lift, a rapid-deployment exercise in Germany, on November 1, 1963. The M60 has a large circle with the number "62" on the turret; the tree hides another similar marking. *National Archives*

Members of 2nd Platoon, Company C, 1st Battalion, 32nd Armor, repair a thrown track during an exercise in West Germany on November 1, 1963. The man squatting to the left is holding a track jack, and another track jack is lying on the track in the foreground. *National Archives*

Crewmen of M60 9B4120 from Company A, 2nd Battalion, 32nd Armor, survey damage to their tank caused by a thrown track in Neustadt, West Germany, on November 4, 1963. The forward mudguard and fender are crumpled, and the rear of the fender is severely damaged. *National Archives*

An M60A1 from the 40th Armor Company is employed in a "Combat in Cities" demonstration in West Berlin on January 22, 1964. The tank is simulating an out-of-commission vehicle that has been stopped by a burning oil slick in front of it. *National Archives*

Private 1st Class Larry L. Cecil, *left*, and Specialist 4th Class William D. McCullough, of the 3rd Armored Division, await assistance after their dozer-equipped M60, registration number 9B5541, broke down during an exercise at Salmünster, West Germany, on September 19. 1965. *National Archives*

Two M60A1 tanks assigned to Company A, 1/73rd Armor Regiment, climb a hill during a field exercise at Camp Roberts, a California National Guard post, on March 24, 1964. The gun-barrel-shaped objects mounted on the main-gun barrels were oxyacetylene gunfire simulators. *Military History Institute*

An M60A1 with Company A, 1/73rd Armor, crosses a small crevice on March 24, 1964, at Camp Roberts, California. M60-series tanks could cross an 8½-foot trench without bridging. The lowered main gun travel lock is visible at the rear of the engine deck. *National Archives*

During Joint Exercise Desert Strike in the Mojave Desert of California in the last half of May 1964, an M60 tank is about to serve as a ferry for a G-758 Willys MD (M38A1) jeep. The tank's turret is traversed to the right, and a gunfire simulator is attached to the 105 mm gun barrel. *National Archives*

At the same location seen in the preceding photo, an M60 tank crosses a river shallows with a Willys MD jeep piggybacked on its engine deck. Another M60 tank is visible in the distance to the right of the nearer tank, and a jeep is mounted on its engine deck as well. *National Archives*

An M60A1 of D Troop, 1st Reconnaissance Squadron, 3rd Armored Cavalry, pauses alongside some ruined buildings at Baumholder, Germany, on June 28, 1964. The crew has applied mud to the white star and the unit markings on the bow to reduce their visibility. *National Archives*

The same M60A1 as in the preceding photo is viewed from a longer distance during an exercise at Baumholder, Germany, on June 28, 1964. The crew also has spread mud on the white recognition star on the turret and registration number on the toolbox, but not on the star on the turret roof. *National Archives*

M60 9B4147, with the number "14" and nickname "BRUTE" on the turret, is aboard a piece of amphibious river-crossing equipment (ARCE) being ferried across the Regnitz River near Bishberg, West Germany, on September 22, 1965. The tank was with 3rd Battalion, 35th Armor. *National Archives*

During the annual Army training tests (ATTs) of 3rd Battalion, 35th Armor, an M60A1 disembarks from an ARCE on the bank of the Regnitz River in West Germany on September 22, 1965. This tank is nicknamed "BOOTLEGGER" and bears the number "23" on the turret. *National Archives*

At Fort Hood, Texas, on July 15, 1968, a member of the 1st Armored Division is crouching next to the turret of an M60, holding a baby. The registration number on the storage box is 9B3130, and on the turret are a white number "007" and a yellow outline of a square. *National Archives*

The crew of an M728 combat engineer vehicle serving with the 26th Engineer Battalion relaxes between assignments at Landing Zone Fat Boy in South Vietnam on November 7, 1968. The barrel and muzzle cover of the 165 mm gun have a darker appearance than the rest of the turret. *National Archives*

An M60A1 attached to Troop C, 14th Cavalry, has taken up a position in a forest north of Fulda, West Germany, in June 1970. Stenciled in black on the cover of the xenon searchlight is the letter "C" over the number "12." A muzzle plug is inserted in the 105 mm gun barrel. *National Archives*

During Phase II of a military exercise called Operation Certain Thrust in mid-October 1970, an M60 assigned to Company A, 63rd Armor Regiment (Aggressors), is crossing an MT-46 light tactical bridge over the Neckar River in the West Germany. *National Archives*

At Fort Hood, Texas, in October 1970, members of the British Royal Hussars are being familiarized with an M60 tank with markings for vehicle 12, Company B, 67th Armor Regiment, 2nd Armored Division. The tube above the main gun is a gunfire simulator. *National Archives*

Among the US Army vehicles paused at an assembly area at Hohenberg, West Germany, while patrolling a stretch of the Czechoslovakian border in the spring of 1970 is an M60A1 tank to the right. Several more M60 tanks are visible in the background. *National Archives*

The crewmen of an M60A1 of Company B, 70th Armor Regiment, 4th Infantry Division, are dressed for cold weather during training maneuvers at Fort Carson, Colorado, around early 1971. The cover for the xenon searchlight has the tank's company letter, "B," marked on it. *National Archives*

The same M60A1 tank shown in the preceding photo, number "24" of B Company, 70th Armor, is maneuvering along a dirt road at Camp Carson, Colorado, around early 1971. The .50-caliber machine gun was not installed in the cupola of this tank. *National Archives*

An M60A1 tank of Company C, 1st Battalion, 72nd Armor, is firing at a moving target at Range No. 7, South Korea, in April 1971. The bore evacuator on the 105 mm gun has a yellow checkerboard scheme over the evacuator's stock Olive Drab color. *National Archives*

M60A1 tanks of Company C, 1st Battalion, 72nd Armor, 2nd Infantry Division, await range-safety clearance at Range No. 7 on the same date as the preceding photo. The tanks have illegible nicknames on the turrets, and the bore evacuators on the 105 mm guns are painted blue. *National Archives*

In another photo in the series taken at Range No. 7 on April 15, 1971, a row of M60A1s of Company C, 72nd Armor, are in firing positions. The nearest tank is registration number 03816669 and has a blue bore evacuator with a yellow band around its center part. *National Archives*

The crews of several M60A1s of Company C, 1st Battalion, 72nd Armor, await orders to commence practice firing at Range No. 7. The closest tank has the nickname "CHALLENGER" in white on the turret. The bore evacuators have red tops and dark-colored bottoms. *National Archives*

This photo and the next several document an M60A2 152 mm gun tank undergoing testing in the Rolling Fork River area at Fort Knox, Kentucky, around 1971. T142 tracks with replaceable track pads are installed on this vehicle. A tow cable is secured to the side of the turret. *Don Moriarty collection*

The commander in the cupola of an M60A2 and the driver of the vehicle survey the terrain ahead. Removable fabric covers hide the coaxial machine-gun port on the left side of the 152 mm gun and the aperture, on the right side of the gun, above the driver's head, for the XM126 gunner's telescopic sight. On top of the mantlet is the box-shaped armored cover for the infrared transmitter. *Don Moriarty collection*

T97 tracks are installed on this M60A1E1 152 mm gun tank, whose 152 mm gun features a bore evacuator. A large quantity of crew baggage and equipment are stashed in and strapped to the turret basket, to the front of which is strapped a spare tank track shoe. *Don Moriarty collection*

An M60A1E1 wends its way over a snowy test course at Fort Knox around 1971. The bore extractor on the 152 mm gun barrel was eliminated on the M60A2 because that vehicle had the closed-breech scavenger system, using compressed air to clear fumes from the chamber. *Don Moriarty collection*

This photo in a series of shots taken at the Rolling Forks range at Fort Knox around 1971 shows an M60A1E2 with its nomenclature stenciled in large, white letters on the side of the turret. The gunner is riding with his upper body exposed on the right side of the turret. *Don Moriarty collection*

A Shillelagh missile has just been fired from an M60A1E2 or M60A2 at a test-firing range. The missile is visible to the front of the smoke emitting from the 152 mm gun barrel. "TEST OPERATION" is stenciled in white on the side of the storage box at the rear of the fender. *TACOM LCMC History Office*

A heavy application of local camouflage in the form of evergreen boughs partly covers an M60A1 during field tests at Fort Hood in February 1972. An oxyacetylene-powered gunfire simulator is mounted on the main-gun barrel. The bore extractor is black, with a white-and-orange sticker. *National Archives*

An "aggressor force" M60A1 with gunfire simulator splashes through mud en route to a defensive position during maneuvers at Fort Hood in early 1972. For the maneuvers, tactical signs with triangles are taped to the turret and glacis, and a yellow "LV" is on the side of the turret. *National Archives*

An M60A1 conducts perimeter defense during field tests at Fort Hood, Texas, in February 1972. Attached to the main-gun barrel is a gunfire simulator, used in training maneuvers. It operated on oxygen and acetylene, fed by hoses from the box-shaped tank on the side of the turret. *National Archives*

M60A1s from the "friendly forces" have been driven back by the "aggressor force" and are retreating along a muddy gully during military maneuvers at Fort Hood, Texas, in February 1972. An indistinct red-and-white design is painted on the xenon searchlight cover of the lead tank. *National Archives*

A member of Company A, 2nd Battalion, 72nd Armor, stacks 105 mm HEAT-TP-T (high explosive, antitank, target-practice, tracer) rounds with blue projectiles at Range No. 7 in South Korea on April 15, 1971. In the right background is an M60 tank. *National Archives*

M60A1 tanks cross a bridge laid by an M60A1 armored-vehicle-launched bridge (AVLB) during a field exercise on July 1, 1972. The front tank has hull markings for the 33rd Armor, 3rd Armored Division, and the cover of the xenon searchlight identifies it as vehicle 16 of Company B. *Defense Visual Information Center*

The commander of an M60A1 leans over the cupola during the filming of *Helping Peace Survive*, an instructional or propaganda film, at Range Nightmare in the Republic of Korea in the spring of 1973. *National Archives*

This close-up view of an M60A1 of Company C, 77th Armor, was taken while it was on a field exercise in April 1974. The tracks are the T97 model. The service headlights are turned on. The housing and handles for the fire-extinguishing system on the upper left corner of the glacis are painted red. A yellow circle with the bridge classification, "50," is also on the glacis. *Defense Visual Information Center*

An M60A1 tank with a dozer blade, attached to 1st Battalion, 11th Infantry, stirs up dust while driving to Firing Ranger 143 during a field problem at Fort Carson, Colorado, on April 17, 1974. The plumbing and reservoir associated with the dozer are visible on the left rear of the vehicle. *National Archives*

As two AH-1 Cobra helicopters hover overhead, the vehicle commanders of a US Army M60A1 unit give hand signals during a field-training exercise on April 1, 1974. On the lead tank, a unit marking "CAV" for cavalry is present, but the number preceding it is illegible. *Defense Visual Information Center*

On the outskirts of Feuchtwangen, Bavaria, West Germany, an M60 has taken up a defensive position behind an embankment during Exercise Reforger '74. The commander, who is standing in the cupola facing to the rear, is wearing a hooded parka to suit the frigid weather. *National Archives*

An M60A1, with another one visible farther down the line, is in a defensive position overlooking Feuchtwangen during Reforger '74. Mounted on the main-gun barrel, to the front of the xenon searchlight, is a Hoffman device, for simulating gunfire during training maneuvers. This vehicle provides an excellent view of the MASSTER camouflage that many 7th Army vehicles in West Germany wore in the 1970s. *National Archives*

During 1976, an M60A2 advances across a field, turret turned toward the right. The unit markings on the bow of this vehicle are indistinct but appear to indicate that the tank belonged to the 1st Battalion, 67th Armored Regiment, 2nd Armored Division. *National Archives*

At Fort Hood, Texas, in August 1975, an M60A1 ascends an embankment alongside a lake. Visible through the lens of the AN/VSS-1(V) xenon searchlight is the lamp housing. Also packed in the searchlight housing were a heat exchanger, blower, igniter, and other equipment. *National Archives*

A heavily dust-caked M60A1 crewed by members of the 5th Battalion, 1st Cavalry, advances toward Jack Mountain during Exercise Brave Shield XII at Fort Hood in late August 1975. The crew has arranged a heavy application of cedar boughs on the tank for extra camouflage. *National Archives*

The presence of a bore evacuator on the 152 mm gun launcher indicates that this vehicle was an M60A1E2 as opposed to an M60A2. Assigned to the 5th Cavalry, it was maneuvering during Exercise Brave Shield XII at Fort Hood, Texas, in late August 1975. *National Archives*

During Exercise Brave Shield X at Fort Carson, Colorado, on November 6, 1974, an M60A1 of the "friendly forces" emits a puff of black engine exhaust as it seeks out enemy forces. The crew of the tank has applied a tan camouflage pattern over the tank's woodland camouflage. *National Archives*

M60A1s proceed along a street in a German city during maneuvers on September 17, 1975. The front tank is operating on the old T97 tracks and is equipped with the old-style xenon searchlight. A low-visibility black recognition star is on the glacis. *Defense Visual Information Center*

Among the tactical vehicles and equipment stored at the motor pool at Fort Bliss, Texas, on September 26, 1975, is a row of M60A1 tanks. The vehicles are painted in the MASSTER camouflage developed by the US Army in the early 1970s. *National Archives*

The first tank in a row of vehicles lined up in September 1975 is an M60A1E2, as distinguished by the bore evacuator on the 152 mm gun barrel and the two forward shock absorbers. *Don Moriarty collection*

Two M60s of the 2nd Battalion, 66th Armor, await their turn to be loaded onto railroad cars at Baumholder in Germany's Rhineland-Palatinate in October 1975. They were destined for Grafenwöhr in Bavaria, where they would be used in support of the troops of Brigade '75. *National Archives*

A column of Patton tanks awaits loading on railway cars at Baumholder, West Germany, en route to Grafenwöhr in October 1975. The closest tank is an M60 marked as vehicle 16, Company C, 2nd Battalion, 66th Armor, 2nd Armored Division. *National Archives*

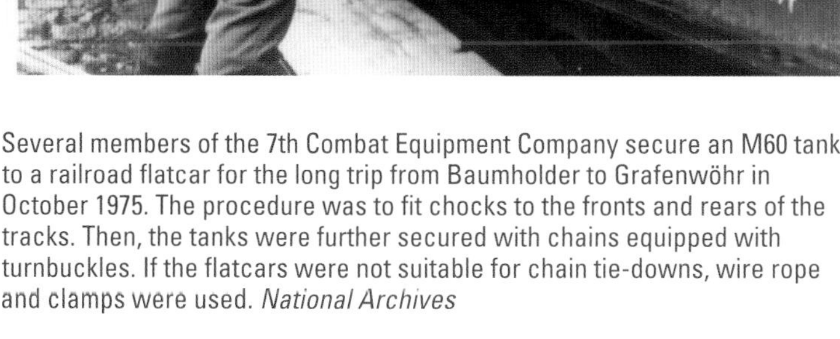

Several members of the 7th Combat Equipment Company secure an M60 tank to a railroad flatcar for the long trip from Baumholder to Grafenwöhr in October 1975. The procedure was to fit chocks to the fronts and rears of the tracks. Then, the tanks were further secured with chains equipped with turnbuckles. If the flatcars were not suitable for chain tie-downs, wire rope and clamps were used. *National Archives*

A column of M60 105 mm gun tanks along a city street wait to be loaded on railroad cars at Baumholder in October 1975. All the turrets are traversed to the rear, and the cupolas are traversed forward, with the vehicle commanders standing up in all the cupolas. Xenon searchlights are installed on the mantlets. Sections of spare tracks are attached to the sides of the turrets of several of the tanks. *National Archives*

The Mobility Equipment Research & Design Command (MERDC) camouflage scheme on this M60A1 on a field exercise at Fort Hood, Texas, in about September 1976 is a good match for the local terrain. The air cleaner on this tank was the unarmored, aluminum, top-loading version. A folded tarp is on the turret roof. *National Archives*

Three M60A1 tanks roll through a populated area during Exercise Reforger '76 in the fall of 1976. These tanks have tactical signs on their glacises: white squares with dark-colored triangles. Unit markings for Company A, 63rd Armor Regiment, are on the bow of the first tank. *National Archives*

An M60A1 tank of the 1st Training Battalion, Company B, Provisional, is used to teach crewmen how to operate and maintain this type of tank, at Fort Riley, Kansas, on November 2, 1976. Here, trainees receive a refresher class on how to drive the M60A1. *National Archives*

An M60A1 tank moves out of a position during a combat-training exercise at an undisclosed desert location on January 1, 1977. The paint is painted in a desert camouflage scheme developed in the 1970s by MERDC. *Defense Visual Information Center*

A dusty M60A1 main battle tank dressed out in local camouflage advances along a dirt road during a field-training exercise at an undisclosed location on April 1, 1977. A large, white number "58" has been painted on the glacis for the exercise. The tracks are the T97 model. *Defense Visual Information Center*

A column of US Army M60A1 tanks move along a dirt road during a field-training exercise in April 1977. These vehicles have a recent improvement: M239 smoke dischargers on the sides of the turrets, with dustcovers over them. Each discharger accepted six smoke grenades. *Defense Visual Information Center*

Members of a North Carolina National Guard armor unit train in M60A1 tanks at Fort Hood, Texas, on June 10, 1977. The third tank in line has its turret traversed to the rear. The tanks have tree branches arranged on them to help them blend in with the foliage and terrain. *National Archives*

Several M60A1s advance during a training maneuver at an unidentified desert location in January 1977. They wear MERDC desert camouflage schemes. The dustcover on the AN/VSS-1 xenon searchlight on the tank to the right has "TK 31" stenciled in white on it. *Defense Visual Information Center*

M60A1s are secured to the deck of USNS *Comet* (T-AK-269) in early August 1977. They were being transported from Bayonne, New Jersey, to Europe, where they would participate in Exercise Reforger '77. The searchlight cover of the tank in the foreground is marked "HQ." *National Archives*

A US Army M60A1 disembarks from the vehicle-landing ship USNS *Comet* (T-AK-269) at the port of Ghent, Belgium, in late August or early September 1977. This tank was among the vehicles and equipment being delivered to northwestern Europe for Exercise Reforger '77. *National Archives*

Turret traversed to the rear, an M60A1 tank is about to drive up the ramp onto an M747 semitrailer hitched to an M746 truck tractor at the maintenance facility at Böblingen, West Germany, in September 1977. This was part of Exercise Reforger '77. *National Archives*

Sergeant 1st Class Nathaniel Telfare, of the Armor School at Fort Knox, invented a mount for a .50-caliber M2 machine gun on the 105 mm gun barrel of the M60A1, as a subcaliber training device. Called the M179 Telfare, an example is seen on an M60A1 at Fort Polk, Louisiana, in 1978. *National Archives*

An M179 Telfare is clamped to the main gun of an M60A1 of the 1st Battalion, 40th Armor, at Fort Polk on May 8, 1978. The Telfare device replicated the ballistic properties of the 105 mm main gun, but at a much-lower cost per round and with more flexibility in suitable firing locations. *National Archives*

Turret traversed to the left, an M60A2 advances into position of a firing range during a firepower demonstration for President Jimmy Carter at Fort Hood, Texas, in June 1978. A few tree branches have been arranged on top of the turret for camouflage purposes. *Defense Visual Information Center*

Members of Company A, 6th Battalion, 68th Armor, wear dust masks as they take an M60A1 to a firing range at Grafenwöhr, West Germany, on July 13, 1978. The commander's and the driver's combat vehicle crewman's helmets are marked with the company letter, A, inside a black circle. *National Archives*

Tankers serving with the 1st Brigade, 84th Division (Training), make field repairs to the forward end of the right suspension of an M60 Patton MBT during a training exercise on the tank course at Fort McCoy, Wisconsin, on July 21, 1978. This tank is fitted with T97 tracks. *National Archives*

An M60A1 with the 24th Infantry Division maneuvers through a pine forest during Exercise Gallant Eagle '79 at Eglin Air Force Base, Florida, in October 1978. It is painted in a woodland MERDC scheme, and a Hoffman device for simulating gunfire is mounted on the main-gun barrel. *Defense Visual Information Center*

The tracks of these two M60A1s barely fit on flatcars at Churchill Dock, Antwerp, Belgium, during Exercise Winter Reforger '79 in January 1979. The nearer M60A1 is registration number 9B8977. Splashes of white have been sprayed on the MERDC scheme, for winter camouflage. *National Archives*

A column of M60A1s, turrets traversed to the rear and main guns seated in travel locks, make their way along a narrow road in southwestern Germany during Exercise Winter Reforger '79 in January 1979. Large wooden crates stenciled "M60A1" are secured to the engine decks. *National Archives*

In late January and early February 1979, a Reforger exercise was conducted in harsh winter conditions in West Germany. Here, an M60A1 negotiates a snowy dirt road during a training exercise. A tactical sign with three vertical bars within a circle is on the turret basket. *National Archives*

An M60A1 assigned to the 1st Battalion, 5th Field Artillery, moves into position during Reforger '79 somewhere in West Germany during late January or early February 1979. A considerable amount of white paint has been applied to the vehicle for winter camouflage purposes. *National Archives*

During Reforger '79 in late January or early February 1979, a crewman standing in the turret of an M60A1 of the 1st Battalion, 63rd Armored Regiment, waves at children along a street in or in the vicinity of Steinbach in south-central West Germany. *National Archives*

An M113 armored personnel carrier and an M60A1 operate side by side on maneuvers during a field-training exercise in May 1979. The M60A1 is equipped with smoke dischargers on the sides of the turret and an AN/VSS-3A infrared and white-light searchlight. *Defense Visual Information Center*

"LADY LOVE" is stenciled in yellow on the dustcover of the AN/VSS-3A searchlight on this M60A1 tank during a field-training exercise at an unidentified location in May 1979. The tank has a weathered MERDC camouflage scheme. The headlights have been removed. *Defense Visual Information Center*

"LADY LOVE" is viewed from a different perspective during training maneuvers in May 1979. On each side of the turret to the rear of the rangefinder was a box containing cartridges for the smoke dischargers. A liquid container is stowed to the rear of that box. *Defense Visual Information Center*

In another view of the M60A1 with "LADY LOVE" stenciled in yellow on the dustcover of the AN/VSS-3A searchlight, a modification to the turret—on the basis of the experience of Israeli M60A1s in the 1973 Yom Kippur War—is apparent. The shot traps created by the overhangs near the bottoms of the sides of the turrets were filled in, resulting in vertical sides at the bottom. To the front of the cupola, next to the hood for the gunner's periscopic sight, is the deflector for the cupola machine gun, made of bent rod and intended to prevent firing at the xenon searchlight. *Defense Visual Information Center*

"LADY LOVE" appears in a final photo as it moves through a forest. The air cleaners on this vehicle (the left one is on the fender below the rear of the turret bustle) are the top-loading aluminum type; the later, armored air cleaners had lifting eyes on the sides. *Defense Visual Information Center*

An M60A1 main battle tank advances through a clearing while on maneuvers during Brave Shield XX, an exercise held at Fort Lewis, Washington, in August 1979. A spare road wheel and a section of two track links are stored on the side of the turret. The muzzle is taped over. *Defense Visual Information Center*

An M60A2 crosses an engineer treadway bridge at an undisclosed location in May 1979. The air cleaner, to the front of the soldier standing next to the center of the tank, is the top-loading aluminum model. A spare road wheel is stowed on the front of the turret basket. *Defense Visual Information Center*

A driver in the Advanced Individual Training Course, US Army Armor School at Fort Knox, Kentucky, is negotiating an M60A1 tank over a rough stretch of road on the Taly Tank Driving Course on May 15, 1979. The .50-caliber machine gun has been removed from the cupola. *National Archives*

This M60A1 tank photographed at the June 1979 Canadian Army Trophy Competition, a NATO gunnery competition in Bergen-Hohne, West Germany, features an interesting pixilated camouflage pattern. This was the Dual-Texture (or Dual-Tex) Digital Camouflage, an experimental scheme employed by the US 2nd Armored Cavalry Regiment in Europe starting in 1978. It was developed by Lt. Col. Timothy R. O'Neill, professor of engineering psychology at the US Military Academy, who determined that a correctly applied digital pattern was more effective in reducing a vehicle's detectability than the NATO and unpatterned camouflage schemes. *National Archives*

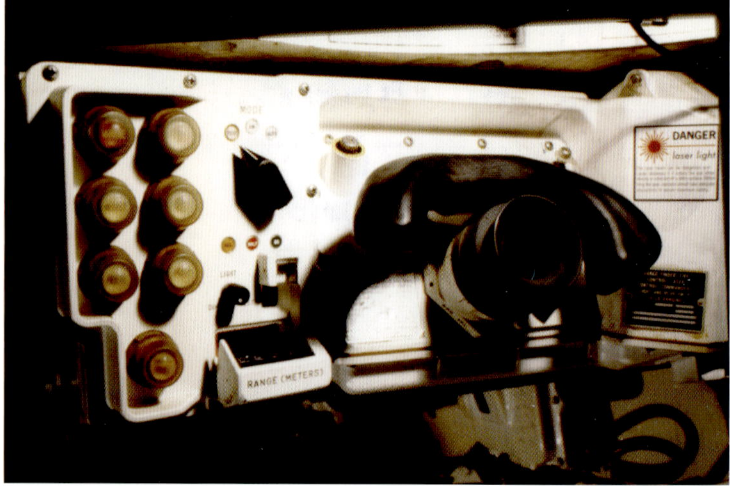

One of the improvements in the M60A3 was the laser rangefinder, replacing the original optical rangefinder. Seen here is the commander's AN/VVG-2 laser rangefinder, including, *left to right*, the commander's control panel, range displays, and eyepiece and crash pad. *National Archives*

SSgt. George Hindle peers through the periscope of his M60A3 during training at Fort Knox in October 1979. *National Archives*

This M728 combat engineer vehicle, registration number 9B8476, moves earth with its dozer blade during an exercise at Fort Belvoir, Virginia, on July 10, 1978. On the dozer blade are the markings "TDC 11" and "SPE 106." There is MERDC camouflage paint on the blade. *National Archives*

The same M728 CEV here has its turret traversed to the front. The crewman to the left, holding a coiled cord for his headset that is running up to the cupola, is communicating with the driver. The .50-caliber machine gun has been removed from the cupola. *National Archives*

An M60A1 tank maneuvers across the McGregor Range at Fort Bliss, Texas, during a training exercise on February 27, 1980. The tank and crew were assigned to the 3rd Armored Cavalry Regiment. A desert MERDC camouflage scheme has been applied to the tank. *National Archives*

The crew of an M60A1 conducts a live-firing exercise at Fort Irwin, California, in March 1980. This tank has the AN/VSS-1 xenon searchlight; the nonarmored, aluminum, top-loading air cleaners; and T142 tracks with replaceable octagonal rubber track shoes. *National Archives*

Members of Company A, 2nd Battalion, 77th Armor, 9th Infantry Division, check to make sure that the M60A1s in this shipment are secured for transport on railroad flatcars at Fort Lewis, Washington, on March 7, 1980. The tanks were about to be taken to the Yakima Firing Center. *National Archives*

Crewmen of Company A, 2nd Battalion, 77th Armor, *to the left*, adjust turnbuckles to secure an M60A1 tank to a railroad flatcar at Fort Lewis on March 7, 1980. Chocks have been fitted against the rear of the tracks and in between pairs of road wheels. *National Archives*

A soldier guides the driver of an M60A1 of the 24th Infantry Division as it disembarks from the vehicle landing ship USNS *Comet* during an Emergency Deployment Readiness Exercise (EDRE) from Fort Stewart, Georgia, to the Port of Savannah, Georgia, in June 1980. *National Archives*

A row of M60A1 tanks of the 156th Armor Regiment are parked in the motor pool at North Fort Polk, Louisiana, on July 31, 1980. At this time, the regiment included four battalions and was incorporated into the 256th Infantry Brigade, Louisiana National Guard. *National Archives*

Viewed from atop the turret during a firing exercise at Schofield Barracks in July 1980, the cupola .50-caliber machine gun fires at a target. Inside the cupola, the white shape above the commander's helmet is the upper part of the M34 daylight periscope mount. The two dark spots on the mount are indentations for wing nuts. *National Archives*

During the same firing exercise at Schofield Barracks depicted in the preceding photo, an M60A1 tank painted in MERDC camouflage conducts a main-gun, live-fire, combat-simulation exercise. It has the late-style armored top-loading air cleaners and an AN/VSS-3A xenon searchlight with a dustcover over it. *National Archives*

The crew of an M60A1, turret traversed to the rear and speeding along a dusty trail at Schofield Barracks, Hawaii, in July 1980, are wearing white dust masks. The vehicle is heavily coated with dust, but here and there, one may see evidence of MERDC camouflage beneath the dust. *National Archives*

A row of M60A1s from 1st Platoon, Company C, 2nd Tank Battalion, 2nd Marine Division, stand ready during Exercise Crisex '81, held in Spain, in October 1981. The rubber tires of the inner road wheels of the closest tank have been severely chewed up by the track guide horns. *Defense Visual Information Center*

A trooper of the 1st Armored Cavalry tosses a grenade at an M60A1 during an urban-combat-training exercise called Doughboy City in West Berlin in November 1981. A small US flag is on the storage box on the fender, and a cover has been placed over the cupola machine gun. *Defense Visual Information Center*

Providing cover for members of the 82nd Airborne Division in Exercise Bright Star in Egypt in December 1981 is an M60A1 tank painted in a sand-colored camouflage scheme. Mounted on the barrel of the 105 mm main gun is a Hoffman device, a gunfire simulator for training. *Defense Visual Information Center*

During a landing exercise at an undisclosed location in December 1981, a Marine Corps M60 disembarks from *LCU-1656*. The turret is traversed to the rear, and a large amount of baggage and equipment is stored in the turret basket and atop the turret. *Defense Visual Information Center*

An M60 tank advances along a snowy forest trail during Exercise Northwind, winter-combat training maneuvers at the North Post Training Area, Fort McCoy, Wisconsin, on February 23, 1982. Fort McCoy was often used as a training area for National Guard armor units. *National Archives*

On April 15, 1982, a demonstration of the comparative mobilities of the M60A3 and the M1 Abrams was held at Aberdeen Proving Ground, Maryland. Here, the M60A3 negotiates a rough course. A man in a white shirt and tie and wearing a CVC helmet is in the cupola. *National Archives*

Members of the 9th Chemical Company, 9th Infantry Division, wash down an M60A1 in an evaluation of the XM17 Sanator lightweight decontamination system at the Yakima Firing Center, Fort Lewis, Washington, in November 1982. *Defense Visual Information Center*

The M60A1 tank being decontaminated at Yakima Firing Center is viewed from another perspective. If the tank came under chemical attack during wartime, it would be crucial to decontaminate it and its crew as soon as possible. The Sanator decontamination unit was man-portable. *Defense Visual Information Center*

An M60A1 assigned to the 2nd Tank Battalion, 2nd Marine Division, maneuvers across Arctic countryside during Exercise Cold Winter '83 in Norway in March 1983. To make the tank blend in with the snowy terrain, white paint has been applied over the green base color. *Defense Visual Information Center*

An M60A1 guards a US Marine encampment on the outskirts of Beirut, Lebanon, in April 1983. The Marines were part of a multinational peacekeeping force sent to Lebanon after fighting between Israeli forces, which had occupied the southern half of Lebanon in 1982, and the Palestine Liberation Organization. *Defense Visual Information Center*

In a photo related to the preceding one, members of an emplaced M60A1 monitor the perimeter of a USMC camp near Beirut in April 1983. An M239 smoke discharger and a smoke-grenade storage box are visible on the side of the turret. To the front of the tank is a sandbagged bunker. *Defense Visual Information Center*

A Marine M60A1 moves into position in a sandy area, well churned up by tracked vehicles, outside a USMC camp near Beirut in April 1983. The tank is fitted with T142 tracks with octagonal rubber track pads. A white band is painted on the 105 mm barrel aft of the fume extractor. *Defense Visual Information Center*

Flying Old Glory from a radio antenna, an M60A1 patrols the perimeter of a Marine encampment on the outskirts of Beirut in April 1983. Above the rear of the left fender is an infantry telephone box, for communicating with the crew from outside the tank. *Defense Visual Information Center*

A Marine M60A1 is situated in a defensive position behind an earthen wall topped with sandbags in front of a house in a suburb of Beirut in April 1983. The tank is painted in a woodland MERDC camouflage scheme, heavy on the greens. Smoke grenades are loaded in the launchers. *Defense Visual Information Center*

A USMC M60A1 tank in woodland camouflage drives off the ramp of a utility landing craft (LCU) and enters the surf en route to join the Multinational Force, to conduct peacekeeping operations in Beirut in May 1983. The rear of the turret is piled high with boxes and equipment. *Defense Visual Information Center*

"PSYCHO" is the nickname stenciled in black on the bore extractor of this Marine M60A1 that has just landed at Beirut as part of the Multinational Force in May 1983. This vehicle has armored, top-loading air cleaners; smoke dischargers; and an exhaust stack for deepwater fording. *Defense Visual Information Center*

Turrets still traversed to the rear and exhaust stacks still mounted after landing, two Marine M60A1s, including "PSYCHO" in the foreground, prepare to move off the beach at Beirut. A compact AN/VSS-3A searchlight is installed on the mantlet. *Defense Visual Information Center*

An M60A3 of 1st Platoon, 48th Brigade, 108th Armored Division, Georgia National Guard, participates in Exercise Company Team Defense at Fort Stewart, Georgia, in July 1983. The cupola machine gun has the Multiple Integrated Laser Engagement System (MILES) barrel, for training. *Defense Visual Information Center*

During Exercise Company Team Defense at Fort Stewart, Georgia, in July 1983, an M60A3 occupies a defensive position. Clearly visible are the two lenses of the gunner's IR/passive periscopic sights, to the front of the cupola. A Hoffman device is on the main-gun barrel. *Defense Visual Information Center*

An M60A3 and M113A1 armored personnel carriers of the 48th Brigade, 108th Armored Division, attack opposing forces during Exercise Company Team Defense at Fort Stewart in July 1983. To the rear of the cupola is the MILES Combat Vehicle Kill Indicator (CVKI) beacon. *Defense Visual Information Center*

Advancing through a recently cleared roadblock during Exercise Company Team Defense at Fort Stewart is an M60A3 of 1st Platoon, 48th Brigade, 108th Armored Division. The detector belt of the MILES suite may be seen running over the mantlet and thence along the side of the turret. *Defense Visual Information Center*

Crew members of the 1st Platoon, 48th Brigade, 108th Armored Division, remove equipment from their M60A3 main battle tanks at the conclusion of Exercise Company Team Defense, at Fort Stewart, Georgia, in July 1983. MILES equipment boxes are in the foreground. *Defense Visual Information Center*

The driver of a US 1st Cavalry Division M60A1 with its turret traversed to the rear is guided along a ramp toward a Federal German army landing craft for transport across the Rhine River at Wesel, West Germany, during Exercise Reforger / Autumn Forge '83. *Defense Visual Information Center*

An M60A3 main battle tank from an unidentified unit crosses a medium girder bridge during Exercise Reforger (Return of Forces to Germany) '83 at Fulda, West Germany, in September 1983. The number "09" is on an orange placard on the side of the turret. *Defense Visual Information Center*

In this elevated view of the town of Fulda, West Germany, during Exercise Reforger '83 in September 1983, an M88 armored recovery vehicle tows an M60A1 or M60A3. Both vehicles have tactical signs on their glacis, a yellow triangle with the number "05" in the center. Fulda was a critical location in the Cold War, being located at the Fulda Gap, a corridor through which the Soviets were expected to attack should full-scale war break out. *Defense Visual Information Center*

US 1st Cavalry Division M60s, including one at the front with an orange "X" on the glacis, cross a ribbon bridge spanning the Maas River in Holland during Reforger / Autumn Forge '83 in September 1983. Combat support boats are working to hold the bridge steady. *Defense Visual Information Center*

M60A3s cross a class 60 medium-girder bridge constructed by the 54th Combat Engineer Battalion, 130th Engineer Brigade, during Exercise Confident Enterprise / Reforger '83 in West Germany in September 1983. Both tanks have yellow placards with the number "11." *Defense Visual Information Center*

Marines man an M60A1 main battle tank in a dug-in defensive position at Beirut International Airport during the peacekeeping mission in Lebanon in December 1983. The airport was in the West Beirut area, and its defense was tasked to the US Marines of the Multinational Force. *Defense Visual Information Center*

Several M60A1s of the 22nd Marine Amphibious Unit are parked inside the Mid East Armor Platoon Headquarters compound while awaiting embarkation on ships of Amphibious Squadron 4 at the conclusion of the Multinational Force's operations in Lebanon in February 1984. *Defense Visual Information Center*

Soldiers of the 2nd Battalion, 25th Infantry Division, advance through the city of Chipyong-Ni during the joint South Korean / US training Exercise Team Spirit '84 in March 1984. Approaching in the background is an M728 combat engineer vehicle with much mud on the dozer blade. *Defense Visual Information Center*

In a photo taken moments from the preceding one, Maj. Edward R. Cruickshank, commander of the 1st Battalion, 299th Infantry, observes the entry of 25th Infantry Division Orange Forces into Chipyong-Ni. In the background is the same M728 seen in the preceding photo. *Defense Visual Information Center*

An M60A1 operated by "friendly forces" moves into the territory held by "opposition forces" during Exercise Air Warrior, at Fort Irwin, California, in March 1984. The vehicle, painted in sand MERDC camouflage, has MILES training equipment and is heavily laden with baggage. *Defense Visual Information Center*

Two M60A1 tanks (*foreground* and *left*), M88A1 armored recovery vehicles, cargo trucks, and other ground-support vehicles are parked in the port staging area of Bremen, northern West Germany, after being offloaded from the vehicle cargo / rapid-response ship USNS *Capella* (T-AKR 293). *Defense Visual Information Center*

An M60A1 has been offloaded onto a wharf at Bremen, West Germany, from USNS *Capella* in June 1984. On the rear of the hull are markings for vehicle 21, Company C, 2nd Battalion, 70th Armor Regiment, 24th Infantry Division. *Defense Visual Information Center*

Private 1st Class John Marquette, a ground guide, stands in front of an M60A1 of Company D, 70th Armor, being offloaded from USNS *Capella* at Los Angeles, California, in August 1984. The tank soon would be part of Exercise Gallant Eagle '84. The tow shackles are painted silver. *Defense Visual Information Center*

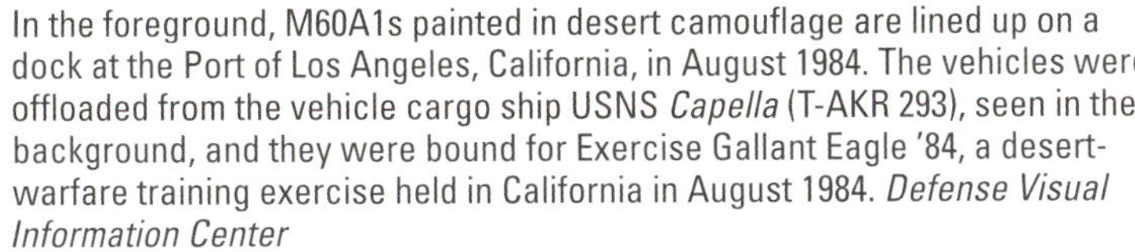

In the foreground, M60A1s painted in desert camouflage are lined up on a dock at the Port of Los Angeles, California, in August 1984. The vehicles were offloaded from the vehicle cargo ship USNS *Capella* (T-AKR 293), seen in the background, and they were bound for Exercise Gallant Eagle '84, a desert-warfare training exercise held in California in August 1984. *Defense Visual Information Center*

A transport ship's crane lowers a sling holding an M60A1 main battle tank onto a railroad flatcar during Exercise Team Spirit at a port in the Republic of Korea on September 28, 1984. This tank has the armored, top-loading air cleaners; smoke dischargers; and smoke-grenade box on the side of the turret, and an AN/VSS-3A searchlight with a cover enclosing it. *Defense Visual Information Center*

Korean dockyard workers positioning an M60A1 on a railroad flatcar for transport to its staging area during Exercise Team Spirit on September 28, 1984. This tank was painted in MERDC camouflage. Soon, the three-color NATO camouflage would replace MERDC. *Defense Visual Information Center*

M60s of the 70th Armor Regiment, including a Company C M60A3 in the foreground, have been loaded on railroad flatcars after being offloaded from USNS *Capella* at Los Angeles. A series of chains and hooks secure each of the tanks to a flatcar. *Defense Visual Information Center*

An M60 main battle tank, equipped with a bulldozer blade, prepares to cross a stream during Exercise Team Spirit '84.

An M60A1 in sand camouflage has thrown a track during a training exercise in 1984. An M88 armored recovery vehicle stands by to assist crewmen in repairing the track. From what is visible of the unit markings, this tank belonged to Company C, 70th Armor, 24th Infantry Division. *Defense Visual Information Center*

An M60A3 main battle tank of Company 3, 32nd Armor Regiment, advances through an unidentified city during Central Guardian, a phase of Exercise Reforger '85, on January 24, 1985. Octagonal orange placards with the number "17" are on the glacis and the turret. *Defense Visual Information Center*

The commander of an M60A3 scans through binoculars for signs of the opposing force during Central Guardian on January 23, 1985. Markings on the bow are for the thirteenth vehicle in the line of march, Company B, 69th Armor Regiment, 8th Infantry Division. *Defense Visual Information Center*

An M60A2 has just exited from the ramp of a US Army LARC-LX (Lighter, Amphibious Resupply, Cargo, 60-ton) on the beach at Fort Eustis, Virginia, on May 13, 1985, during Prolog '85, a US Army Transportation and Aviation Logistics exposition. *Defense Visual Information Center*

An M60A3 main battle tank plows through a gap in an embankment in the California desert during Exercise Gallant Eagle '86, on July 27, 1986. MILES equipment is installed on the tank, for gunnery training, including a Hoffman device and a Combat Vehicle Kill Indicator (CVKI). *Defense Visual Information Center*

During an armored-assault drill in Exercise Gallant Eagle '86, the turret of an M60A3, probably the same one seen in the preceding photo, is trained on a potential target to the left. On a bracket on the rear of the right fender is an infantry telephone box. *Defense Visual Information Center*

An M60A1 of Company A, 1st Tank Battalion, attached to the 1st Battalion, 2nd Marine Regiment, occupies a partially concealed position during Exercise Kernel Potlatch 86-1. The company letter, "A," and the vehicle's order of march, "25," are stenciled on the searchlight cover. *Defense Visual Information Center*

A pair of mechanized landing craft assigned to Assault Craft Unit 2 (ACU 2) ferry two M60A1s to the amphibious transport dock USS *Trenton* (LPD-14) in the Atlantic during the field-training Exercise Solid Shield '87 in January 1987. The tank on the left is equipped with a dozer blade. *Defense Visual Information Center*

An M60A1 leads the pack during field maneuvers. These tanks have MERDC camouflage, named after the developer of the scheme, the US Mobility Equipment Research & Design Command. The scheme features two shades of green, black, and white. *Department of Defense*

A force of M60A1 main battle tanks advance during maneuvers in a field-training exercise. All of the tanks have the smaller-sized AN/VSS-3A infrared searchlights, which replaced the bulkier AN/VSS-1(V) xenon units, mounted over the mantlets. *Department of Defense*

A US M60A1 wades ashore at a harbor in the Republic of Korea during Exercise Team Spirit '88 in March 1988. A large, wooden crate is strapped to the top of the armored air cleaner; one of the straps is attached to the outboard lifting eye on the side of the air cleaner. *Defense Visual Information Center*

Turrets piled high with crates and gear, two M60A1 main battle tanks pause on a floating causeway before coming ashore during the joint US / South Korean Exercise Team Spirit '88. The maritime pre-positioning ship USNS *Sgt. William R. Button* (T-AK 3011) is anchored offshore. *Defense Visual Information Center*

A USMC M60A1 has been secured to the deck of a landing craft, air cushion (LCAC), following an amphibious warfare demonstration at San Diego, California, in March 1989. The tank is equipped with bolted-on reactive armor, which explodes on contact to defeat hollow-charge projectiles. *Defense Visual Information Center*

A Marine Corps M60A1 main battle tank with the reactive-armor suite attached to the hull and the turret and a Hoffman device clamped to the main gun advances after coming ashore in North Carolina on D-day of the joint-services exercise Solid Shield '89 in May 1989. *Defense Visual Information Center*

Utility landing craft *LCU-1634* ferries two USMC M60A1s equipped with reactive armor, along with a truck and a bulldozer, during the combined Thai/US joint Exercise Thalay Thai '89 in September 1989. Tarpaulins are arranged over the cupolas of both tanks. *Defense Visual Information Center*

Three members of the crew of an Egyptian M60A3 pose next to their tank during Operation Desert Shield, the 1990–91 buildup to Operation Desert Storm, the campaign to eject the Iraqis from Kuwait. Camouflage netting is arranged over the rears of the hull and the turret. *Defense Visual Information Center*

A Marine M60A1 with reactive armor and an M9 bulldozer surmounts a sand berm on Hill 231 in Saudi Arabia in January 1991. The crew of the tank, from Company D, 2nd Tank Battalion, were rehearsing their role in Task Force Breach Alpha in the kickoff to Operation Desert Storm. *Defense Visual Information Center*

The same M60A1 with M9 bulldozer kit seen in the preceding photo has moved a few feet forward while breaching a berm on Hill 231 in January 1991. The tank is painted overall in sand or tan, with a roughly sprayed-on black "V" marking on the reactive armor of the turret. *Defense Visual Information Center*

In February 1991, an M60A1 of the 2nd Marine Expeditionary Force, equipped with reactive armor, a mine-clearing plow, and an orange recognition panel, prepares to lead AAVP7A1 amphibious assault vehicles into Kuwait at the start of the ground phase of Operation Desert Storm. *Defense Visual Information Center*

A Marine Corps M60A1 with reactive armor and painted in woodland camouflage is parked in an area adjacent to US Navy Fleet Hospital 5 in Saudi Arabia during Operation Desert Storm, in February 1991. Green tape has been wrapped over the muzzle to keep out the elements. *Defense Visual Information Center*

A Marine makes an adjustment inside the cupola of an M60A1 parked in a maintenance area in northern Saudi Arabia during Operation Desert Storm, on February 1, 1991. The arrangement of reactive armor attached to the hull and turret is evident. *Defense Visual Information Center*

A Marine crewman of an M60A1 grabs a few winks on a cot next to his tank while other crewmen to the right work on the power pack in northern Saudi Arabia on February 5, 1991. The nickname "LEFTY" is stenciled in black on the fume extractor of the 105 mm gun. *Defense Visual Information Center*

The same M60A1 shown in the preceding photo, with the crewman napping next to it, is viewed from a different angle. In addition to the reactive armor on the tank, the crew have piled sandbags on the glacis and at points on the turret for an extra measure of protection. *Defense Visual Information Center*

While work on the power pack of the M60A1 continues at a camp in northern Saudi Arabia on February 5, 1991, Chief Warrant Officer 2 Ed Bailey, *left*, a Navy Reserve photographer, confers with a Marine. The engine deck has been placed on the ground behind Bailey. *Defense Visual Information Center*

Marines continue to work on a power pack that has been removed from the M60A1 in the background on February 5, 1991. The heart of the power pack was the Continental AVDS engine, linked to the Cross-drive CD-850-6A transmission and the final drive. *Defense Visual Information Center*

In the final scene from the series taken at a camp in northern Saudi Arabia on February 5, 1991, a fuller view is available of the power pack that has been removed from the Marine M60A1 tank in the background. The rear of the transmission is in focus. *Defense Visual Information Center*

Members of the 72nd Engineering Company, 24th Infantry Division, test a mine-clearing rake attached to an M728 combat engineer vehicle during Operation Desert Storm. The rakes appear to be attached to the standard M9 dozer blade. *Defense Visual Information Center*

With the blade and plow lowered, the M728 advances toward the cameraman. *Defense Visual Information Center*

In the midst of oil wells, blazing after the retreat of Iraqi forces, a Marine Corps M60A1, an M88 armored recovery vehicle, and two M998 HMMWVs are dispersed along a road in Kuwait during Operation Desert Storm on February 27, 1991. *Defense Visual Information Center*

As the sun sets on February 27, 1991, US Marine tankers perform maintenance on their M60A1 main battle tanks during the pursuit of Iraqi forces retreating from Kuwait. A mine-clearing plow and a mine roller, detached from the tanks, lie on the ground between the two vehicles. *Defense Visual Information Center*

An Egyptian army M60A1 tank takes part in a live-fire session during Exercise Bright Star '94 on November 18, 1993. A roll of camouflage netting tied to the turret partly obscures the Arabic numbers painted on the metal. On the side of the turret is a marking consisting of a red triangle over a green bar. *Defense Visual Information Center*

During Exercise Bright Star '94, an Egyptian M60A1 passes along the side of a line of M1 Abrams main battle tanks on November 18, 1993. The main gun is in the full-recoil position. The xenon searchlight and the smoke dischargers are not mounted on this vehicle. *Defense Visual Information Center*

General Data				
Model	**M60**	**M60A1**	**M60A2**	**M60A3**
Weight*	102,000 lbs.	105,000 lbs.	114,400 lbs.	114,600 lbs.
Length**	366.5	371.5	286.85	371.5
Width**	143	143	143	143
Height**	126.34	128.23	130.31	130.31
Track	T97E2	T97E2	T97E2	T97E2
Track width**	28	28	28	28
Crew	4	4	4	4
Maximum speed	30	30	30	30
Fuel capacity	375	375	375	375
Range	250	310	250	280
Turning radius	pivot	pivot	pivot	pivot
Armament				
Main	105 mm M68	105 mm M68	152 mm M162	105 mm M68
Secondary	1 × .50 cal. M85	1 × .50 cal. M85	1 × .50 cal. M85	1 × .50 cal. M85
Coaxial	1 × .7.62 mm M73	1 × 7.62 mm M73 or M219	1 × 7.62 mm M73/M73A1 or M219	1 × 7.62 mm M219
Ammunition				
Main	57 rds., 105 mm	63 rds., 105 mm	46 rds., 152 mm	63 rds., 105 mm
Secondary	900 rds., .50 cal.	900 rds, .50 cal.	1,080 rds., .50 cal.	900 rds., .50 cal.
Coaxial	5,950 rds., 7.62 mm	5,950 rds., 7.62 mm	5,500 rds., 7.62 mm	5,950 rds., 7.62 mm

* Fighting weight

** Overall dimensions listed in inches. Measure with main gun facing forward, and antiaircraft machine gun mounted

Engine Data	
Engine make/model	Continental AVDS-1790-2
Number of cylinders	90-degree V-12
Cubic-inch displacement	1,791
Horsepower	750 @ 2,400
Torque	1,710 @ 1,800
Governed speed (rpm)	2,400

The Infantería de Marina (Marine Infantry of the Spanish Navy) obtained a number of M60A3 main battle tanks, such as this one, which is coming ashore at al-'Umayyid, west of Alexandria, Egypt, on October 20, 2001, during Exercise Bright Star '01/'02, joint US-Spanish-Egyptian amphibious maneuvers. *Defense Visual Information Center*

An M60A3 from Spain's Infantería de Marina churns up sand as it advances across a beach during a landing exercise at al-'Umayyid, Egypt, on October 20, 2001. Although difficult to discern, "INFANTERÍA DE MARINA" is painted in black on the forward storage box on the fender. *Defense Visual Information Center*

In January 2010, members of the 233rd Transportation Company, based at Fort Knox, Kentucky, work to unload an M60A1 or M60A3 tank from a heavy-equipment transporter to one of the post's tank ranges. Here, the tank would serve as a target for a new generation of tankers. *Defense Visual Information Center*

Obsolete, war-weary, or damaged-beyond-repair tanks sometimes end their careers as firing-range targets, to hone the skills of aerial and ground gunners. Such was the case of the M60A3 to the right at an unidentified firing range in 2006. *To the left*, another tank has just been shot up. *Defense Visual Information Center*

In the 1990s, the Florida Fish and Wildlife Conservation Commission conducted a program of constructing marine artificial reefs from man-made materials. The FWCC secured quantities of decommissioned M60s, M60A1s, and M60A3s for the purpose, some of which are on a barge bound for a site. *Florida FWC Artificial Reef Program*

A forklift pushes an M60 overboard from a barge. "Reef-X" is spray-painted freestyle in white on the fender storage box. The vehicle was painted in the long-discontinued MERDC woodland camouflage, and the engine compartment door-grilles had been removed. *Florida FWC Artificial Reef Program*

Another M60 has been pushed overboard from a barge, to serve as part of a man-made reef off the Florida coast. In the background, more M60A1 tanks await their turn to plunge into the deep. The Defense Logistics Agency arranged the transfer of the tanks to the artificial-reef project. *Florida FWC Artificial Reef Program*

A diver investigates an M60A1 main battle tank resting on the floor of the ocean, part of the artificial reefs constructed in the 1990s. In addition to M60A1 and M60A3 MBTs, the reef project also made use of M113 APCs and M551 Sheridan armored-reconnaissance/airborne-assault vehicles. *Florida FWC Artificial Reef Program*

The chassis of M60A1 AVLB, USMC registration number 557940, of the 1st Combat Engineer Battalion recovers its bridge at Combat Outpost Ouellette in Afghanistan on February 16, 2011. The M60A1 AVLB was among the last US military uses of the M60 series. *Gunnery Sgt. Bryce Piper / USMC*